What They Knew and When They Knew It

How the 3M Company Knowingly Contaminated the World

by
Thomas G. Anderson

<u>DEDICATION</u>

*To the concerned, informed and caring people
who helped write this work:*

Becky Berg *who asked that question*

Lori Swanson *for being someone who still
carries a passion for the cause*

Kevin P. Murphy, PhD. *– The best possible
advisor for this project*

And my Executive Editorial Committee:

Mackenzie*, my granddaughter*

Sara*, my daughter*

and

Kelsey*, my wife and partner throughout our
3M journey and life beyond*

Table of Contents

PROLOGUE

My Story, My

Confession

"They're going to plant a bomb in your car," said my wife.

"I'm not revealing anything that isn't already a part of the public record," was my measured response.

"What about your pension?"

That's the tone the conversation takes when you tell your spouse you are about to document how your former employer of thirty-six years polluted the global ecosystem and tainted the bloodstream of nearly every human being on Earth in a span only slightly longer than your own lifetime. Weighing the options, I thought I had made the right decision; in my heart I knew she agreed but her pragmatism, tinged with more than a little fear, was perfectly justifiable. It would be a tale of David and Goliath. The big money machine versus a lone retiree. The nameless cog set against the overpowering corporate machine. Hallways full of attorneys and spin doctors teamed up against me. The odds, as they usually are in such stories, were not in my favor; in real life rarely does the little guy win. This is a deeply personal story that at the same time encompasses the globe. It started in my own backyard and seeped into the collective bloodstream of the planet. We are united by a toxic contaminate first manufactured by the company that issued me a paycheck for nearly four decades. Someone once said "my honor is my loyalty" but, in this instance, my loyalty is tempered by my honor.

Her words came back to haunt me at the strangest times. One windy and icy December evening I gingerly made my way across a glistening and slippery parking lot; it was, after all, Minnesota in the days before Christmas. There, sitting solitary in the filtered moonlight was my car, an island of Detroit-made metal surrounded by a sea of empty parking spaces. My hand grasped its key in my pocket. Soon it would be in the ignition but then what? Boom or a ride home? Why had she said that? Why had she planted that seed? Too many Grisham novels, perhaps, made her think this might be a case of life imitating art. Why would a gigantic corporation be intimidated by a lone retiree armed with the truth gleaned from the pages of public records? They wouldn't. They would not be intimidated by me nor I by them. This was a tale of truth not fiction; their own words would form the narrative. My job would be to assemble the story more than write it, copy-and-paste rather than create.

She had been there through the entire journey. We had been so proud to be a part of a global, dynamic company. Pride swelled even more when we – I say we because my wife, my kids and me were all honored to be a part of this warm, corporate family – were transferred from Michigan to the corporate headquarters in Minnesota. The work family here was tens of thousands. Everywhere we went the name was recognized and revered. Everyone knew the company. Everyone knew the products. Everyone thought of you as something special since you worked for such a legendary corporate entity. They paid for me to finish my college education.

They paid for my family to enjoy the company country club. There were recognition dinners, river cruises, golf outings, picnics and concerts provided by the company. Provided by the company that all the while was poisoning the world. Like a corporate Jekyll and Hyde, it dished out hamburgers and hot dogs while polluting the very water we drank and ignoring the facts from both internal and external sources of the damage it was causing. For many, many years they knew they were manufacturing a toxic chemical, using it in everyday products that they sold to an unsuspecting public and yet chose to do nothing but continue churning out those products that turned into profits in the billions of dollars.

They knew and they kept the awful truth to themselves. Like four of a kind, they kept their cards close to their corporate chest. They had conducted numerous tests on the product. The lab animals had died: fish, mice, rats, goats, monkeys. The workers who manufactured the chemical had elevated levels of the contaminate in their blood. They even moved women of child-bearing age out of the production facilities knowing full well the chemicals damaged the fetuses in other mammals. Over the years they continued to monitor those production workers by collecting data and sera samples, but the chemical always showed up; it never seemed to go away. They came to the conclusion early on the chemical didn't break down or disperse on its own; it even earned the nickname "the forever chemical" from chemists themselves.

Doctors began calling it "bio-persistent" and "bio-accumulative". Bioaccumulation happens when you absorb a substance faster than your body can get rid of it. Bio-persistence means that substance, once it's in your body, has a hard time leaving. The combination of those two traits indicates you are absorbing a substance and just never, ever fully ridding yourself of it. It was said "you may not die from it, but you will die with it".

Two years before I joined the company, they were alerted by two outside researchers that the chemical was showing up in blood samples from people not just in Minnesota where the product was being manufactured, but from multiple locations across the United States. The company denied any responsibility, "plead ignorance" as they described it in an internal memo at the time, but later conducted their own tests on blood samples and came to the same conclusion. Polar bears at the North Pole have the contaminate in their blood, penguins at the South Pole, fish in the Mississippi River, your family pet, and the sacred waters of the Ganges. When further tests were conducted on blood samples, the first instances where researchers found non-contaminated blood were from a collection of Army inductees during the Korean War. Nearly every blood sample tested for this substance since then has shown at the very least trace amounts. Newborn infants have tested positive as have samples from every human ever tested on every continent. All of this happened and the company, my company, did nothing to remediate the problem, withdraw the chemical from the marketplace or even offer a humble apology. Their objective became managing the problem rather than solving the problem. Hard-hearted, uncaring and profit-driven they continued for decades manufacturing a substance they knew to be dangerous, toxic and by their own admission carcinogenic with no remorse or concern for their fellow human beings or the very globe we jointly inhabit. They shamelessly caused irreparable harm to the planet and every living substance on it.

As employees we were told none of this, they just served us another hot dog and continued to poison the planet. We were unaware of the skeleton in the closet down the hall from our office in the guise of a wonderful product with a toxic ingredient, or the crazy uncle authorizing tests on animals that ended with them killing all animal participants even the ones in the control groups or of the criminal cousin who tried for years and years to keep the truth buried and the profits rolling in; the family secret was well hidden. Then the shadows began to lengthen, questions were raised, innocent, unsuspecting neighbors became ill, some were dying. Through testing, the state of Minnesota found well water and aquafers were poisoned particularly and most dangerously around four specific sites where the company had just dumped the product and its waste onto open ground in unlined pits and trenches. Levels spiked around the company's plant where the chemicals were manufactured. The state found out. The state sued. The state won that suit. In the pantheon of environmental litigation, it is Deepwater Horizon first, the Exxon-Valdez second, and the settlement awarded the State of Minnesota third. Added together nearly one- billion-dollars has been award to be used exclusively by the nine communities affected most, not nearby cities, not rural Minnesota or neighboring states, just for the roughly 162,300 people living in those nine cities, towns and townships many of whom are company employees and retirees. Divided evenly, the original award means over $5,200 for every man, woman and child living in the affected area; some argue that's not nearly enough for a lifetime of drinking contaminated water.

"Are you from Woodbury?" Ironically, my water bottle, a gift for my volunteerism from the city where I lived, had revealed my identity.

I had decided to go back to college to earn a second Bachelor's degree and perhaps the lofty goal of a doctorate under a state-supported Senior Citizen

Education Program. The class was "Public History" with a specific focus that semester on environmental justice. The question had come from a classmate. Our project for the term was to propose a museum exhibit based on a local environmental justice issue of our choosing. Do well enough and we would earn a display at the state's museum of natural history.

"What do you think about doing something around PFAS contamination in South Washington County?" she asked. It had come full circle. I was retired but my employer was calling to me once again. This call was not to come back to work but to tell the awful truth about what they had done to the global community. I simply answered "yes"; the time had come to take a stand.

How do you do that? How can you expose the wrongdoings of an organization that over the years kept a roof over your head, paid for your food, health care, even at one point gave you a car to drive? The answer? It was very easy. They made it easy by their ill-formed strategies, deception and deceit. Telling this tale would be among the most difficult things I would ever do but remaining silent was not an option. No one was telling this tale and it needed to be told. There will be those among former and current employees who will consider me a heretic, a turncoat. While there will be many who see this as biting the hand that fed me, there is pride in knowing I wasn't one who drank the Kool-Aid® of corporate corruption nor was I swayed by spin. The friends I lose among former fellow employees for not falling in line like an unquestioning child will be replaced tenfold by a whole new set of people who see evil in a company that touts its own environmental stewardship while, in reality, irresponsibly polluting the entire planet. An organization like that isn't one you can say you are proud to have been a part of, or to support, or respect.

Asking for perfection, especially of a corporation, is foolhardy. Excellence, however, is attainable and should be a never-ending goal. Excellence in this case was often applied to the means and methods of hiding the facts. They have become adept at standing between the whole truth and an unsuspecting and unaware public. Nothing to see here, move along.

From that singular moment in the classroom we moved forward, aided, abetted, encouraged and supported by a professor who saw what we were doing as important work. We earned plaudits, high praise and A's but there was more to this story than what could be told on a museum display panel, so much more. "You have a unique perspective on this" the professor told me at semester's end.

When I mentioned the name I would be naming would be my former employer he assured me "that's the story and it needs to be told". We decided that night to do a deeper dive into the subject and I would spend an entire semester doing an independent study around this particular aspect: the company and its years of hiding the truth. The university's motto is "Driven to Discover"® and with that I felt a new sense of being emboldened, empowered and overwhelmed by the obligation to tell this story. It is eye-opening, frightening and worst of all an affirmation that dastardly deeds by corporate America are not just fodder for the movies; they do, indeed, happen in real life.

This is a tale of my former employer of thirty-six years and how mistake piled on mistake has cast a new light on a company that I was once proud to be a part of. Make no mistake: they caused obscene and irreversible damage to the planet and every living organism that inhabits it and covered it up. They caused obscene and irreversible damage to the planet and every living organism that inhabits it and denied culpability. They caused obscene and irreversible damage to the planet and every living organism that inhabits it and withheld critical information from the responsible governmental agencies and the public for decades. What follows are facts, not corporate hyperbole.   Based upon the preponderance of evidence presented here, you can decide for yourself.

<u>CHAPTER ONE</u>

*From the Very
Beginning There Was
Deception – A Quick
Chronology*

"It is almost always the cover-up rather than the event that causes trouble" – Howard Baker

It is a story that lacks public outrage. Despite newspaper articles, television reports, documentaries and Hollywood films it has lingered beneath the radar. The finest spin doctors available have deftly deflected attention away from this story and its global consequences. Years of deceit, misdirection and corporate corruption kept it from piquing public interest.

Not even litigation in excess of a billion -- with a "b" -- dollars made PFAS a household name. Though it may not be a household name, it poignantly permeates every household, and, to date, nearly every living creature ever tested.

This first chapter will briefly highlight key events in the PFAS saga. These are merely the touchstone moments in this environmental tragedy that will serve as an introduction to this tale. These moments and this chapter, for the most part, will serve as the outline of what follows. The purpose of these first few pages is to tell you what is going to be told in greater detail in subsequent chapters. This tragedy, this case of global environmental justice, has numerous plots, storylines and tentacles but the purpose of this work is a focus on the corporate corruption that was the underlying cause. It wasn't the chemical itself that is to blame, it is the company that created it and their decision to put profit before environmental well-being. Evil wrongdoing by Fortune 500 companies is not just the stuff of Hollywood films or best-selling novels; it happens in life every day and this case of evil wrongdoing encompasses the world. And the one telling the tale is a person whose resume lists thirty-six years of experience with that very same company. In a sense, it is a voice from the inside.

The last few years of the 1940's and the first few of the 1950's the world was PFAS-free for the last time. In the period since, that family of substances has gone from non-existent to universally present but not in a good way. PFAS is a fluorocarbon substance constructed by bonding fluorine and carbon molecules together. Carbon is one of the most abundant elements on planet Earth, but its PFAS companion, fluorine, is a dangerous, insidious substance that some chemists refer to as "the Devil's piss". Chapter Three will delve into PFAS chemistry itself but for now just realize it is a compound unlike any other compound. As will be mentioned later, more than one chemist has called PFAS "insidious".

In conjunction with each other and when applied to the surface of an item, PFAS performs the useful purpose of preventing liquids from seeping into said items. PFAS does that by lowering the surface tension of the item that it is applied to – let's say it's your raincoat – PFAS is then acting as a surfactant preventing your raincoat from sopping up all that rainwater by repelling it. Whenever you don't want a liquid to infiltrate an object, cover that object with a PFAS-based compound.

It didn't take long to see that PFAS could be used on a variety of items from the aforementioned raincoats and outdoor equipment as a waterproofing additive to furniture and carpet as a stain repellent to cookware as a non-stick surface. The products used for those applications became household names: Scotchgard®, Teflon®, Stainmaster®. The world fell in love with these PFAS-treated items. It was a newly found expectation that our muffins wouldn't stick to the muffin tin, the wine spilled on the carpet would bead up and wait to be wiped away, that the soda pop the kids splattered on the sofa wouldn't penetrate the fabric and with a swipe of a sponge your furniture was as good as new. Our lives were better and easier. One of the company's involved even told us it was "better living through chemistry". But the story being told here is not one of better living but rather the dark, deceitful, sinister side of corporate corruption.

The company that gave us better living through chemistry was E. I. du Pont de Nemours and Company known globally as the chemical giant DuPont. Around since the early 1800's, the DuPont name may not be as familiar as its portfolio of products: neoprene, nylon, Mylar, Kevlar, Tyvek, Lycra and Teflon® to name a few. It was Teflon® that DuPont used as the brand name for their non-stick cookware coating first introduced to homemakers in 1962. That coating used PFAS as a critical ingredient. At one point, Teflon® was even used to rustproof the interior of the Statue of Liberty.

Teflon® was accidently created by DuPont in 1938 using a compound called PTFE. World War II and specifically the Manhattan Project found a use for PTFE. So impenetrable was this substance that it was used to coat valves and seals in pipes containing a highly volatile form of uranium used in the enrichment process when making the atomic bomb.

Make no mistake in thinking that DuPont acted alone in hiding the dangers of PFAS. It had a co-

conspirator and that partner in deception was 3M Company, my former employer. DuPont saw the value of a non-stick substance and wanted to manufacture the coating under the brand name Teflon®. But there were issues. They couldn't make mass batches of the product with any consistency. It would clump up, one batch would be just fine, the next useless.

Enter 3M. At 3M, scientists had been working with fluorocarbons as well and had devised a method of mass producing the substance through an electrochemical process. 3M applied for and was granted a patent and a deal was struck between the two companies: no need for both to build manufacturing plants to make PFOA, one of the early configurations of PFAS, 3M would build the facility and sell PFAS to DuPont. Thus, the poisonous 3M/DuPont partnership was born.

From those days in the early 1950's until 2002, DuPont never once made a single ounce of PFOA. Yes, they used PFOA in the manufacturing process, but they bought all of the PFOA ever used in every Teflon® coated item for over fifty years from 3M. It is 3M's PFOA that permeates the planet, but the mutual and even cooperative corruption started back on day one of their relationship.

Ramping up for increased PFAS production, 3M built a manufacturing facility intended for the manufacture of the substance utilizing their newly patented process in Cottage Grove, Minnesota. Named after the first farm platted there, Cottage Grove boasted a population of only 884 people in 1950[1] so the decision

---

by 3M to build a massive manufacturing facility on over 1,700 acres in the community was a major coup. Construction began in 1947 and the first production of PFAS began less than two years later. Sitting on a picturesque portion of the upper Mississippi River, the Cottage Grove manufacturing location would serve not only as a pastoral backdrop for the massive complex but as a convenient means for disposing industrial waste into that legendary waterway, America's premier river.

Shortly after production began -- along with the disposal of the waste and by-product from the production of PFAS -- 3M plant workers began noticing environmental oddities in and around the Cottage Grove plant. There seemed, for example, to be a lot more dead fish on the banks of the Mississippi. In 1958 a team of investigators from the Minnesota Department of Health conducted tests at Cottage Grove. Specifically, what they tested was effluent, the water discharged back into the Mississippi River after it had been treated by 3M in their newly constructed wastewater treatment facility at the plant. In the MDH's words, "to test the toxicity of the wastes on fish life"[2]. They submersed fish taken from the river into a tank of effluent. The fish suffocated to death in less than five minutes. When the effluent was oxygenated, the fish lasted 96 hours before dying. Even after being presented with the test results, nothing other than keeping a watchful eye on the nearby iconic waterway was done as a result of this early rudimentary

---

https://www2.census.gov/library/publications/decennial/1950/population-volume-1/vol-01-26.pdf Page 25

[2] Exhibit 1020, State of Minnesota v. 3M Co., Court File No. 27

test; no reporting, no further tests, no public announcement.

Two years before, in 1956, production and its inevitable waste had increased to the point where a secondary disposal location was needed in addition to the one on-site at the plant. Twelve miles north of Cottage Grove sat the more rural than suburban community of Oakdale. An existing commercial dumpsite became the next disposal location for the "liquid and solid industrial waste"[3] from the Cottage Grove manufacturing plant and none of the roughly 3,200 Oakdale residents knew anything about the lurking danger.

Four years later the Oakdale site was full to, literally, overflowing with waste including PFAS residue. Less than five miles north of the Cottage Grove plant, now called Chemolite, was a 240-acre tract of land in what was then known as Woodbury Township. 3M sent out an assessment team from their corporate geology department. Going into the acquisition 3M had "no plans to construct a drainage pond or to provide burning facilities"[4] meaning the plan was to just dump the waste

---

[3] Exhibit 1872, State of Minnesota v. 3M Co., Court File No. 27
[4] Exhibit
1025,
State of
Minnesot
a v. 3M

directly on top of the ground. They knew then this was not a "good neighbor" location. They knew they wouldn't be welcomed to the neighborhood with open arms. Their internal report read, in part: The zoning problem alone merits a good deal of attention. Low level land usage such as dumps, and gravel pits generally brings condemnation from nearby residents and population centers.[5]

Yet the land was acquired, Oakdale was shut down, and all the previously mentioned "liquid and solid industrial waste" would be dumped in Woodbury.

As the sixties ebbed into the seventies and as dumping in the Twin Cities continued to expand, a pair of independent researchers contacted 3M in 1975. These two doctors had begun seeing traces of PFAS in human blood samples. 3M had known PFAS had been leeching into the groundwater around those SoWashCo

---

Co., Court File No. 27 [5] Exhibit 1025, State of Minnesota v. 3M Co., Court File No. 27

dumpsites since the early 1960's. It didn't take much scientific imagination to think that PFAS ingested through well water would find its way into a person's bloodstream. These samples, however, were not from South Washington County, Minnesota. They were from thousands of miles away in places like Texas and New York. Less than twenty-five years after its first full-scale production, PFAS had now spread incredible and incomprehensible distances from quiet Cottage Grove.

In a confidential 3M document recounting the August 14, 1975 conversation between those researchers and a 3M scientist, the writer says of the outside researchers: "Somewhere he got the information that 3M's fluorocarbon carboxylic acids (what we now call PFAS) are used as surfactants and wanted to know if they were present in 'Scotchgard' or other items in general use by the public."[5] The writer then, almost boastfully says, "We plead ignorance"[6] knowing full well the answer to the question asked by the researcher was "yes". Plainly put, he lied. Soon, blood samples collected by 3M would show the problem was far more widespread than just a few other states. Despite finding their chemical in human blood, 3M said nothing, reported nothing to the proper authorities but kept making PFAS and profits from its sale. It continued to be business as usual, business now being conducted under an ominous dark cloud.

It is Thursday July 21, 1983 and an article in the *New York Times* carries the headline "Minnesota Mining to Clean Up Waste Dump". The article explains how 3M is going to spend more than $6 million cleaning up the Oakdale dumpsite. "Although the dumps in Oakdale are

---

[5] Exhibit 1118, State of Minnesota v. 3M Co., Court File No. 27
[6] Exhibit 1118, State of Minnesota v. 3M Co., Court File No. 27

considered less threatening to public health and the environment"[7] says the article at one point adding later the dump contains "such toxic chemicals as benzene and ethyl benzene"[8] with no mention of fluorocarbons, or PFAS.

This latest episode had begun the preceding September when 3M itself found 11,500 cubic yards of waste in deteriorating barrels and drums at the Abresch site in Oakdale, 11,800 tons of which had been transported directly from the plant in Cottage Grove, obviously containing at least some waste from PFAS manufacturing. 3M did not disclose that in their report of the find to the EPA. Nearby wells – thirty-nine in all -- were contaminated and abandoned. The EPA declared all three 3M dumpsites in Oakdale (Abresch, Brockman and Eberle) as meeting the qualifications for the Comprehensive Environmental Response, Compensation, and Liability Act of 1980 (CERCLA) known throughout environmental circles as Superfund. Collected as a tax on the chemical and petroleum industries and set aside as a trust, CERCLA established "prohibitions and requirements concerning closed and abandoned hazardous waste sites and provided for liability of persons responsible for releases of hazardous waste at these sites".[9]

Clean-up of the sites was completed in 1984 and after the final grading was completed this once toxic substance disposal area was repurposed as open space

---

[7] *New York Times*, MINNESOTA MINING TO CLEAN UP WASTE DUMP, July 21, 1983
[8] *New York Times*, MINNESOTA MINING TO CLEAN UP WASTE DUMP, July 21, 1983
[9] https://www.epa.gov/superfund/superfund-cercla-overview

for all to enjoy. The Abresch site today is split in two by a major highway with over 12,000 vehicles passing by every day. It is surrounded by housing developments, businesses and retail shopping. You can get your car washed, have lunch and work all day adjacent to a Superfund clean-up site. My daughter's office is on the northeast edge of the Abresch site; looking out her office window she has a view of the remains of a EPA Superfund cleanup site. That is the legacy of the 3M Oakdale dump.

As a superfund condition, the land is tested by the EPA every five years. The last test revealed "the remedy is functioning as intended and is protective of human health and the environment in the short term. Long-term protectiveness will be ensured once institutional controls are in place" which is EPA-speak for the chemical residue will be there for a long, long time. 3M global headquarters is about four miles to the southwest of the former Oakdale dump site; it might as well be in another universe since still buried at that site are the evils committed by 3M long ago that they would much prefer Oakdale and the rest of the world forget.

The Occupational Safety and Health Administration (OSHA) was created "to ensure safe and healthful working conditions"[10] for all Americans. They monitor working conditions and, among many responsibilities, serve as a mechanism to warn employers and employees of potentially dangerous conditions, processes or products. According to the OSHA website, a Material Safety Data Sheet (MSDS) is a safety document required by the Occupational Safety and Health Administration (OSHA) that contains data about the physical properties of a particular hazardous

---

[10] https://www.osha.gov/aboutosha

substance. MSDS sheets are created for a variety of hazard materials including compressed gases, flammable and combustible liquids, oxidizing materials, poisonous or infectious material, corrosive material and dangerously reactive materials.[11]

Issuing an MSDS is pretty routine for most products where an element of worker or user safety is in play. They state the facts about the product under required topics such as ingredients, environmental information, suggested first aid and the like. When you purchase a product that requires an MSDS those details are automatically sent every time you buy that item: buy the product and an MSDS is sure to follow as sure as day follows night. It was a normal course of business, then, that on February 7, 1997 3M sent DuPont an MSDS on "FLUORAD BRAND Fluorochemical Surfactant"; in other words, PFAS. Under section 8 is a standard reporting requirement called "Health Hazard Data". Right after a section on "If Swallowed" and before information titled "Mutagenicity" is the heading "CANCER".

Underneath that title on that MSDS sheet from 1997 are these words:

> ***WARNING: Contains a chemical which can cause cancer. (3825-26-1) (1983 and 1993 studies conducted jointly by 3M and DuPont).***[12]

---

[11] https://msdsauthoring.com/msds-safety-data-sheet-chemicals-osha-msds-rules

[12] Exhibit 1471, State of Minnesota v. 3M Co., Court File No. 27-CV-10-28862

If it had been encircled by neon lights it could not have stood out more on the page. There, in black-and-white and in their own words was an admission that both 3M and DuPont had determined the PFAS they were manufacturing and using as a raw material in consumer goods was carcinogenic. And if that wasn't enough of an astonishing admission of guilt, the MSDS also clearly spells out that both companies not only knew of PFAS's cancer causing properties but knew of them fourteen years prior to the publishing of that MSDS sheet. In short, through tests they themselves conducted, 3M knew PFAS to be a cancer-causing substance as early as 1983 and reaffirmed their findings in 1993. It was now 1997 – 43 years after they killed fish with PFAS, 41 years after they began dumping PFAS waste in Oakdale, 37 years after doing the same in Woodbury, nearly thirteen years after Oakdale was declared an EPA Superfund Cleanup site, and close to fifteen years after conducting tests that concluded PFAS to be carcinogenic – and it would be another year before they would start to admit their lies and deception and the long and wearying journey to the truth would begin. Never once, of course, has 3M owned to the cover-up.

As late as 2008, when "the University of Minnesota released a new study of death rates among PFOA workers at 3M. The findings: elevated rates of stroke and prostate cancer. 3M's response will sound familiar: 'Nothing in this study changes our conclusion that there are no adverse effects from PFOA'".[13]

---

[13] Robert Bilott, *Exposure: Poisoned Water, Corporate Greed and One Lawyer's Twenty-Year Battle Against DuPont* (New York, Atria Books, 2019) pg. 287

Rob Bilott, a Cincinnati-based corporate defense attorney who filed and won a PFAS based class action lawsuit against DuPont asks: "What was 3M hiding behind the thick smear of public relations gobbledygook?"[14]

Full disclosure never seemed to be a consideration for 3M. Through all the testing of water, of humans, of vegetation, of fish, mice, rats, eaglets, goats and monkeys the company continued to hide the results from the public, the public that consumed PFAS laden goods in massive quantities. The tipping point happened in 1998 fully forty-eight years after 3M had been granted U.S. Patent No. 2,519,983 for the "Electrochemical Process of Making Fluorine Containing Carbon-Compounds" in 1950. Forty-eight years of manufacturing PFAS. Nearly all that amount of time discovering, testing and realizing the negative affects PFAS had on the planet and its inhabitants. And once discovered, knowingly keeping the facts from those being contaminated.

What follows is a story of corporate corruption on the part of 3M, how they covered up facts and created a self-serving narrative. The topics mentioned above will be the markers along the road of revelation that tell the tale of 3M Company's involvement with PFAS proliferation. We'll explore the companies, the chemistry, animal tests, disposal, proliferation but most importantly the conspiratorial nature of how 3M never full disclosed the nature and dangers of this product for almost fifty years. Nearly all of the information contained herein is part of the discovery process for a suit brought by the State of Minnesota against 3M on

---

[14] Robert Bilott, *Exposure: Poisoned Water, Corporate Greed and One Lawyer's Twenty-Year Battle Against DuPont* (New York, Atria Books, 2019) pg. 52

behalf of the nine communities in South Washington County (SoWashCo) most severely affected by the callous disposal of industrial waste by 3M. These are, for the most part, 3M's own words. These are all public records. This, for the first time, is full disclosure and transparency about what 3M did and didn't do, what they knew and when they knew it.

CHAPTER TWO

A Brief History of

3M

"3M had so many opportunities to do the right thing, but they never did" – Lori Swanson, former Minnesota State Attorney General

It is important to look at a brief history of 3M before we go much further. There had never been so much as a ripple of scandal in the company's history. The only whiff of previous impropriety came in 1975 when, during the Watergate investigation, it was uncovered that three 3M executives had made illegal political contributions to the Nixon re-election campaign. One of the trio was then CEO Harry Heltzer. He and the company were charged with misdemeanors. Heltzer pleaded guilty and resigned from his position. Even with that blip on the radar screen of dishonesty, 3M had always made lists of "Most Admired Companies" or "Best Places to Work". Despite a global presence it always seemed to reflect its Minnesota roots which makes the swerve to deception in the case of PFAS contamination so puzzling. That switch, that alter ego, is hard for some to believe or understand. Perhaps that is the root cause of the lack of public outrage about not only what 3M did to the globe but the lengths they went to in trying to cover up what they knew and when they knew it.

3M, much the same as DuPont, is more commonly known for the products it produces instead of its corporate name. Post-it® Notes, Scotch® Magic™ Tape, Command® Adhesive, Littman™ Stethoscopes are a mere handful of the brands that contribute to the 60,000-item product portfolio. The diversity is also reflective in the span of awards on the corporate mantle that range from an Oscar® to a profile in the Tom Peters best-seller "In Search of Excellence". Considering its near disastrous start still being in business seems a major accomplishment in itself.

"By the shores of Gitche Gumee, by the shining Big-Sea-Water[15]" stands the small town of Two Harbors, Minnesota. It was there in 1902, with the waves of Lake Superior lapping the rocky shoreline, the original Minnesota Mining and Manufacturing Company – soon to be referred to as "3M" and eventually changing its legal name to that acronym -- was founded on a mistake. The name is a clue as to the original aim of the five founding businessmen: from the North Shore of **Minnesota** they were going to begin **Mining** a mineral called corundum and convert it into a viable consumer product by **Manufacturing** sandpaper. Minnesota Mining and Manufacturing was born with one slight birth defect – it wasn't a corundum mine they owned but a mine full of anorthosite, a piece of geology worthless as a component in sandpaper. After that first batch of anthracite the company would mine no more. Instead they would relocate to Duluth where they would purchase the raw materials needed to manufacture sandpaper, but it took fourteen years before the company was able to pay a dividend to its stockholders.

After sinking hundreds of thousands of dollars into the fledgling company, St. Paul investor Lucius Pond Ordway was growing weary of trying to watch over his business investment in faraway Duluth. In 1910 he paid to move the company to St. Paul where he could keep a

---

[15] Henry Wadsworth Longfellow, *Song of Hiawatha*, (Boston, Ticknor & Fields, 1855)

closer watch over things. He also helped build a state-of-the-art sandpaper manufacturing plant in the new location and had the foresight to also retain a cost accountant named William L. McKnight. Those two – Ordway and McKnight – share the title of "founding father of 3M"; they both righted the company's course in its early years, both served as President and McKnight would go on to be Chairman of the Board of 3M until his retirement from the company in 1966 and served as Honorary Chairman until 1972. Those names linger today throughout the Twin Cities with the Ordway Center for the Performing Arts in downtown St. Paul and McKnight Road that forms the western boundary of the 3M Corporate Campus along with charitable foundations in both their names.

William L. McKnight

In 1948, McKnight laid out what many consider to be the very foundation of 3M Company when he said: As our business grows, it becomes increasingly necessary to delegate responsibility and to encourage men and women to exercise their initiative. This requires considerable tolerance. Those men and women, to whom we delegate authority and responsibility, if they are good people, are going to want to do their jobs in their own way. Mistakes will be made. But if a person is essentially right, the mistakes he or she makes are not as serious in the long run as the mistakes management will make if it undertakes to tell those in authority exactly how they must do their jobs. Management that is destructively critical when mistakes are made kills initiative. And it's essential that we have many people with initiative if we are to continue to grow.[16]

Notice that the word mistake(s) is mentioned four times in that short paragraph. Keep that word in mind as you continue to read. The mistakes McKnight envisioned were probably nowhere near the cataclysmic ones made by the company as it dealt with the volatility of PFAS.

The growth of the auto industry coincided with the founding and early years of 3m. Being in the sandpaper trade meant they did business with the plethora of automakers and auto body repair shops that proliferated during the 1920's. Beginning in the '20's two-tone paint jobs became all the rage. Witnessing firsthand the frustration of painters who saw their two-tone paint jobs ruined by the strong bond tape and butcher paper that was the standard of the time, 3M scientist Richard Drew combined a lower tack adhesive with a one inch wide roll of crepe paper and masking tape was born; sort of. Product tests were conducted in a variety of paint shops in cities across the Midwest. The adhesive on the original version of the tape had a tendency to migrate toward the center of the strip of crepe paper leaving the edges "unsticky". When Drew checked back with one of the painters testing the original product to get his reaction the painter reportedly said "this is a great idea but you have to take

---

[16] https://en.wikipedia.org/wiki/William_L._McKnight

it back to your Scotch (implying "cheap") bosses and tell them to put more glue on it."

The derogatory name, unlike the original tape, stuck and another product – Scotch® Brand Tape – was born because of a mistake. During the height of the PFAS era the company launched what many would argue is their iconic product: Post-it® Notes, another product founded on a mistake. By that time 3M was a company known for, among many products, its wide variety of adhesives; adhesives that stick. This new one didn't. Spencer Silver, the man who created the actual Post-it® adhesive, had been looking to create a "super glue" type product but he got the chemical composition wrong and instead created an adhesive that didn't fully adhere because only a small fraction of the adhesive structure was actually touching the surface of the item it was placed upon. In reality, Post-it® adhesive is very strong; that's what gives it its removability and repositionability. An adhesive that doesn't stick isn't really an adhesive; it's worthless. Not this one. It would be worth billions of dollars in sales as it would turn out, but no one early on saw that coming. Realizing that he might have made another in a long line of 3M missteps, Spen put the jar of adhesive on the shelf and moved on.

Among the legends of 3M is the "15% rule". There was a time when employees were encouraged – almost required – to spend 15% of their time on something totally unrelated to their job responsibilities. If you were an accountant and wanted to fiddle around with a new type of packaging tape you were allowed to do so and use as many of the company resources as possible. Secretaries working on new sales reporting systems, chemists working on a marketing project;

nothing was considered out of the realm of possibility. It all seemed to date back to Mr. McKnight and his idea of unleashing employees and allowing them to show initiative. That idea seemed to fade away during my tenure and so it was that the structure of the company seemed to grow more than the desire for individual initiative.

Spen Silver worked alongside fellow scientist Art Fry. Art spent his Sunday mornings singing in the choir at church. He marked that day's selections with strips of paper, but they kept falling out of his hymnal not only causing Art to lose track of what hymns were next, but also littering the choir stall with little pieces of paper. It probably wasn't divine intervention but one day Art remembered Spencer Silver's non-sticky adhesive and dabbed a dollop on the back of a scrap of paper. It stuck to the pages of his hymnal yet removed easily without damaging the paper itself and it could be repositioned and removed many times over. That was how an adhesive that didn't stick became a universally used product. That was the mistake that gave birth to the Post-it® Note.

Those are endearing mistakes, if you will. A mine full of useless rock that led to the creation of one of the largest selling brands of sandpaper in the world. An insult hurled at a roll of tape that created a product name that has become a generic term: Scotch® Tape. And the adhesive that doesn't stick; over and over and over again. Those are the stories we, as employees used to tell. Self-deprecating tales of a rags-to-riches company, of an organization that relished and never criticized its internal mistakes became a part of the daily lives we employees felt comfortable with.

In a way, those are "cute" mistakes which are so unlike the sinister mistake exemplified by the chain of events that deceived the world and, more importantly, delayed for decades humanity's ability to deal with the problem of PFAS proliferation sooner and perhaps even provide a solution. It will be difficult for some to understand this episode given the company's respectable past. Many employees, retirees and extended family never will accept the facts and historical events from the '50's, '60's, '70's, '80's, '90's and on into the 21st Century. They will believe the company spin and the company's hallowed past. Then there will be those who sit on the fence of wanting to still believe in the goodness of this corporate icon. Hopefully the final group, the group that will embark on this journey of discovery with an open mind and realize the tragedy 3M has wrought not only on itself but mankind. Again, bear in mind that what follows is historical fact based on 3M documents and public records.

It is time to turn the page and discover a bleaker chapter in the history of 3M.

<u>CHAPTER THREE</u>

The Chemistry of
PFAS for the Non-Chemist

"Anything with the 'F' (the Periodic Table of Elements symbol for Fluorine) in it is something you want to be very wary of."[17] *- Glenn Evers, Former DuPont Chemist*

---

[17] *The Devil We Know*, Stephanie Soechtig, Jeremy Seifert - Directors, Joshua Kunau – Producer, 2018, YouTube.

As a chemical compound, PFAS is a complex subject. To start with, dispel the notion that PFAS is a singular item, a solitary thing. PFAS is the collective name for what is estimated to be over 4,000 different variations. Much the same as we say "cars", there are Fords,

Chevys, Subarus, etc. PFAS, then, is the collective name for variations like PFOA, PFOS and PFBS. What this section should do is avoid the complex mumbo-jumbo techno speak of chemistry and explain what PFAS does rather than its complex construction. The goal then is akin to the old saying: we're going to tell you the time, not tell you how to build a watch.

PFAS is an acronym which stands for perfluoroalkyl substances and polyfluoroalkyl substances. Simplifying the difference between per- and poly-, without getting too technical, all of the carbon atoms in the per- are totally fluorinated while at least one atom in the poly- is not. To abide by the creed of simplicity pledged earlier, we'll just refer to PFAS as a category.

Danger levels persist in both per- and poly- so there is no need to differentiate their toxicity.

The "P" for per- and poly- are almost always followed by the "F" for fluoro- as in fluorocarbon. If there is fluoro in the name it is a substance that has, as its base, the element fluorine. Fluorine has been called the most reactive of all the elements. It swings wildly from being used in common, humanly helpful items like the fluoride in your toothpaste and drinking water to its dangerous inclusion in PFAS compounds. How it reacts and if it becomes friend or foe depends a great deal on what other elements or compounds fluorine is combined with and how they are combined.

PFAS is made up of a chain of links of carbon and fluorine atoms. Because the carbonfluorine bond is one of the strongest, these chemicals do not degrade in the environment causing it to be bio-persistent. Scientists have been unable to estimate the environmental half-life for PFAS, which is the amount of time it takes 50% of the chemical to disappear. Noble minds have been unable to tell how long PFAS chains stay in the environment.

When 3M received a patent in 1950 it was for an electrochemical process of taking fluorine through a series of electro and chemical steps and ending up with a useable fluorocarbon on a massive scale. In other words, this process took something highly reactive – fluorine – and combined it with something abundant – carbon – and created mass quantities of the compound which became the cornerstone of PFAS.

Why use fluorine? Because, quite simply, fluorine creates very strong bonds so when combined with carbon to create fluorocarbons they bond together to form an incredibly strong barrier. And they also drastically reduce surface tension which makes them a surfactant. It is that quality that makes them an especially effective firefighting agent.

The simplest way of defining why PFAS is used in consumer applications is, because of that incredibly strong and tight chemical bond, it stops liquids from migrating from one place to another. Take the 3M product Scotchgard®. Spray Scotchgard® on your carpet, furniture or clothes. It prevents liquids like juice, wine or mud from penetrating the fibers of the treated item because it has, in essence, coated those fibers in an impenetrable coating of PFAS-based liquid. Same with a

microwave popcorn bag. That bag is just paper. Coat
the inside of that paper bag with PFAS and you keep the
hot oil in the bag and not all over the inside of your
microwave. Dental floss is just string. Coat it with PFAS
and your saliva won't penetrate the string and render it
painful and useless.

The two most common variations of PFAS are PFOA
whose compositions are depicted above. PFOA was the
version of PFAS found in Telfon®. PFOS was the key
ingredient in Scotchgard®. They are similar but different,
yet their one commonality is they both belong to the
PFAS family and they are also the two most abundant
versions of PFAS found in human sera and the
environment. Of the thousands of variations of PFAS
and their global presence, PFOA and PFOS top the list of
most prolific.

        In 2000 when 3M announced its phase-out of
PFOA and PFAS production, the search began for a safer
alternative. That alternative is called Gen-X, but it is by
no means safer than its predecessors. Gen-X chemicals
have been found in surface water, groundwater, bottled
drinking water, rainwater, and air emissions in some
areas just like its older versions. Instead of the old
recipe, Gen-X uses HFPO dimer acid and its ammonium
salt instead of perfluorooctanoic acid (PFOA). Whereas
old school PFAS was an eight-chain construction (eight or
more carbon atoms in a molecule) Gen-X is a six-chain
construction. And it would appear the issues of bio-
accumulation and bio-persistence remain with the new
product they just might not be "forever" chemicals.

        The issue with PFAS is its use in a variety of
products that make life easier rendering PFAS an almost
necessary evil. Looking at some of the consumer

products that are or have used PFAS brings that point home in dramatic fashion.

Takeout Food Containers

Did you ever why wonder the grease from the pizza delivered to your home doesn't seep outside the box? The answer is PFAS. Remember, PFAS prevents the migration of liquids from one place to another so if you coat the inside of a pizza box with PFAS, the grease stays in the box and doesn't leak out onto your car seat, kitchen counter or dining room table. Same thing with the wrapper around your takeout sandwich or burger; the inside of the wrapper is PFAS coated to prevent your "special sauce" from leaking.

Non-stick Pots, Pans and Utensils

All that needs to be mentioned here is the brand name Teflon®. What helped create the coating that prevents things from sticking to an item coated with Teflon® is PFOA or, as DuPont likes to refer to it, C-8. Same thing. Until recently all fluorocarbon substances were constructed using an eight-chain carbon/fluorine structure: C (Carbon) 8 (as in eight chain construction). Teflon® is also used as a coating on some irons, parchment baking papers, sandwich and waffle makers, some ironing board covers, ski bindings, hair straighteners and curling irons.

Microwave Popcorn Bags

This is a category in and of itself. The PFAS prevents the oil, which becomes hot oil in your microwave, from leaking through what is essentially a paper bag, and coating the inside of your microwave with a greasy layer of a hard-to-clean substance. There are some PFAS-free microwave popcorn products on the market including Newman's Own® and Jolly Time®.

Stain- or Water-repellant Clothing

If water beads up on your jacket or coat, it has been treated with PFAS; you have enclosed yourself with PFAS. You are left with a decision: can I live with a wet jacket or can I live with PFAS? One or the other.

Outdoor Equipment

It is difficult to find a tent or sleeping bag that isn't water resistant. After all, what good is a tent that doesn't repel rain or a sleeping bag that sops up water like a sponge? The same as your jacket or coat, more than likely if your sleeping bag and tent are water repellant, they're coated with PFAS.

<u>Stain Treatments for Furniture</u>

If your furniture bears a tag that says "stain resistant" the manufacturer has almost certainly used a PFAS-based product to treat it. The market is now shifting away from proudly announcing "Treated with Scotchgard®" or "Treated with Stainmaster®" to a new type of tag which reads "Treated with a PFAS-free stain repellent".

Carpeting and Carpet Treatments

The relationship between carpeting and PFAS is fairly simple to explain. Nylon carpeting comprises about 95% of all carpet sales while wool, the more expensive alternative, captures the remaining 5% of installations. If your carpet is of nylon construction, it has been treated with a PFAS-based stain repellant. Wool carpeting is never treated for stain resistance by the manufacturer because wool has lanolin which acts as a natural stain repellant. The odds are usually pretty good that, home or away, you have a 95% chance of walking barefoot across a PFAS treated floor. If you treat your carpet or rugs with a retail stain repellant like Scotchgard®, you're treating it with PFAS.

<u>**Certain Cosmetics**</u>

A recent Dutch study found that nearly one-third of the cosmetics they tested contained elevated levels of PFAS-related substances. Shampoo and hairspray of course, but levels exceeding acceptable amounts were found in foundations and concealers. Most were Teflon® based products. There are PFAS-free cosmetics, so the old adage is true: check the label before buying to make sure you're not applying PFAS directly to your face. Also, some shampoos contain PFAS as an additive to help repel airborne contaminants.

Dental Floss

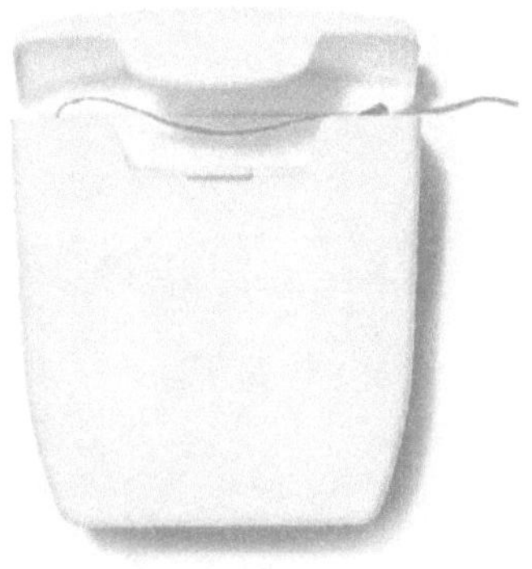

Dental floss? Why would dental floss have PFAS in it? To make it slipperier is the short answer. Coating string with a PFAS-based product turns basic string into an easier to use product. Here's the catch: some dental floss brands, including some of the best-sellers, have PFAS while others don't. Once more, read the label; if there's any mention of fluorine or fluorocarbon there will be PFAS content. But PFAS in dental floss is becoming a moving target so a quick search of the internet will tell you if your brand is among those containing PFAS.

<u>Fire Fighting Foam</u>

       Unless otherwise noted on the canister itself, all foam-based fire extinguishers contain PFAS. AFFF fire-fighting products are used where there is a potential of a flammable liquid fire; these are sometimes called Class B fires. If you see a fire extinguisher with a bold letter B on it, that extinguisher is foam based and, in all likelihood, that foam contains PFAS. A later chapter will deal exclusively with AFFF, aqueous film forming foam.

Water

Forty-nine out of fifty states have PFAS tainted groundwater. The lone exception at the time of this writing is Hawai'i. The latest estimate is when over a third of all Americans turn on the tap, the water that flows from the faucet is contaminated with PFAS. Whether the water source is surface water – water that originates from a nearby river or lake – or groundwater from an underground aquafer, it is PFAS laced. There are treatment options such as reverse osmosis and granular activated carbon (GAC) but reverse osmosis works under high pressure to remove particles while GAC filters need to be replaced frequently. Neither remove 100% of PFAS in water but can remove enough to bring it down to acceptable levels. Either type of filter can remove enough PFAS to make your drinking water safe.

Some bottled waters are reverse osmosis filtered or GAC filtered but very few are both. Again, read the water bottle label to be sure since it's best to have water that has gone through <u>both</u> types of filtration systems. It is also important to remind you that PFAS is a global issue so it might be a false assumption to think a water from a source outside the United States may be safe; there is no guarantee of that. It's also important to remember it's not where the water comes from but if it is filtered before it is bottled. To be safe, read the label to make certain your bottled water has been commercially filtered.

CHAPTER FOUR

*Tests Conducted,
Results Ignored: The Long
List of Lab Animals Who
Died of PFAS Poisoning*

"The greatness of a nation and its moral progress can be judged by the way its animals are treated." - Mahatma Gandhi

It was September 1958. Dwight D. Eisenhower was midway through the first of his two terms as President of the United States, the *Donna Reed Show* was about to premiere on ABC television, and "Volare (Nel Blu dipinto di blu)" by Domenico Modugno was the number one song in the land. Cars had fins, telephones had cords (and the phones remained inside the home, office, or phone booth) and the Minnesota Department of Health had been called to the 3M Chemolite plant in Cottage Grove to analyze how the company was disposing of liquid waste at the site. It was early autumn on the banks of the Mississippi in the Upper Midwest. Fishing season would soon give way to hunting season but in the next few days some fish from the river would meet an unexpected and unfortunate end.

The purpose of the inspection trip by the Health Department was three-fold: to analyze the quality in addition to the quantity of waste being discharged into the Mississippi River after treatment at the plant, and to "determine the adequacy of the waste treatment facilities now in use"[18]. Jump ahead to the "Summary" section of the report written as a follow-up to that inspection visit: "Additional waste treatment or control facilities are needed"[19]. In other words, the waste treatment facilities, built after being granted a permit by the Water Pollution

Control Commission a little more than three years earlier in August of 1955, was already inadequate and improperly treated waste, including residue from the manufacture of PFAS chemicals, was being pumped into the Mississippi River. Decades later it would be determined that it wasn't the inefficiency of the waste treatment plant that was at fault, it was the chemicals being treated, PFAS – the "Forever Chemical" – that was the issue. No technology available at the time could have removed PFAS from that water; no technology available at the time could have prevented PFAS pollution of the Mississippi River.

Amongst the data collected by the Minnesota Department of Health in the autumn of that year were results of tests on effluent, the term used for water as it exits a waste treatment facility. The test subjects? "A number of test fish native to the Mississippi River in this area"[20]. Some of the test results and observations are unnerving. Statements like:

---

[18] Exhibit 1020, State of Minnesota v. 3M Co., Court File No. 27-CV-10-28862
[19] Exhibit 1020, State of Minnesota v. 3M Co., Court File No. 27-CV-10-28862
[20] Exhibit 1020, State of Minnesota v. 3M Co., Court File No. 27-CV-10-28862

> *"...the protective mucous of the fish became altered such that over longer periods of time it might seriously impair the essential processes of the fish or make it more susceptible to diseases which could lead to its death"*[21] .

Or "after short periods of time in the ... waste, the fish appeared to lose their sense of stability"[22]. The following finding seems particularly disturbing especially when you think in terms of other fish and/or mammals:

> *"During the testing, it was also noted that the pectoral and pelvic fins and the gill regions of the fish became unusually red, indicating either a physiological response to or an irritation from the waste".*[23]

Keeping in mind effluent is defined as "an outflowing of water or gas to a natural body of water, from a structure such as a wastewater treatment plant"[24] that effluent should have already been treated and, as much as possible, returned to its original state. In the case of Cottage Grove, the effluent is water that should be ready to be deposited directly back into the Mississippi River. The words of the Minnesota Department of Health in their summary should have

---

[21] Exhibit 1020, State of Minnesota v. 3M Co., Court File No. 27-CV-10-28862

[22] Exhibit 1020, State of Minnesota v. 3M Co., Court File No. 27-CV-10-28862

[23] Exhibit 1020, State of Minnesota v. 3M Co., Court File No. 27-CV-10-28862

[24] https://en.wikipedia.org/wiki/Effluent

raised a red flag when the report stated: "in the undiluted effluent, the test fish died within 3 to 5 minutes".[25] The report continued "if a source of oxygen was introduced into the waste, the fish were able to survive for more than 96 hours"[26].

It should have been obvious from these reports that something was depriving the water of its oxygen and, therefore, suffocating the fish. What is even more worrisome is neither the Health Department nor 3M knew what they were dealing with and neither had the intellectual curiosity to explore these oddities, these abnormalities any further. Had they done something then who knows where we might be now. The fish dying should have sounded an alarm bell, but it didn't. It could have led to early exploration of a cause and cure, but it didn't. It was 1958 and already the Mississippi River was being contaminated with effluent that didn't contain enough oxygen to sustain marine life. It would be another forty years before the truth emerged and the fish that died in 1958 would be counted among the animal victims; there were many more to come. "If a substance shows toxic effects in animals, you must assume it is likely toxic to humans until proven otherwise"[27]

Around 1963, 3M scientists, in conjunction with the United States Navy, found that PFAS, when used as a base in a liquid foam construction, had excellent fire-fighting abilities. The product, the generic term being

---

[25] Exhibit 1020, State of Minnesota v. 3M Co., Court File No. 27-CV-10-28862
[26] Exhibit 1020, State of Minnesota v. 3M Co., Court File No. 27-CV-10-28862
[27] Robert Bilott, *Exposure: Poisoned Water, Corporate Greed and One Lawyer's Twenty-Year Battle Against DuPont* (New York, Atria Books, 2019) pg. 172

Aqueous Film Forming Foam or A-triple F for short, literally formed a film on the fuel in a fire and smothered it. Again, PFAS plays a unique and critical role in the AFFF saga and Chapter Ten is dedicated to that story. In a test of AFFF by a company called Chemical Concentrates Corporation the same thing happened to some unsuspecting fish. 3M patented its version of AFFF in the early 1960's and called it Light Water®. Chemical Concentrates Corporation used Light Water® for their test and "(t)he effects were highly derogatory to marine life"[28]

Those tests, conducted in 1970 and reported to the National Fire Protection Association in June of that year, used Goldfish, Black Moors (a variety of tropical fish) and Calico bass. "At all concentrations ... erratic motion, loss of stability, and other visibly odd effects were present"[29]. The report ends with these final conclusions: "two principal causes of death ... (t)he erratic motion, rapid rotation and general inability to remain upright led to the apparent drowning. There also appears to be an attack of the nervous system as evidenced by high speed swimming and crashing headlong into the sides and bottom of the tank"[30] an obviously horrible way for anything, including fish, to die. This was another test, this time from an outside independent source, that showed PFAS-based chemicals were sucking the oxygen out of water. The condition of low levels of oxygen is called hypoxia and fish react to hypoxia just like you or I would: by going on a desperate

---

[28] Exhibit 1083, State of Minnesota v. 3M Co., Court File No. 27-CV-10-28862

[29] Exhibit 1083, State of Minnesota v. 3M Co., Court File No. 27-CV-10-28862

[30] Exhibit 1083, State of Minnesota v. 3M Co., Court File No. 27-CV-10-28862

search for oxygen in order to stay alive. They would exhibit all the signs of desperation displayed in the 1970 Chemical Concentrations Corporation test.

The 1958 test results were not the first that showed the dangers to animals associated with the PFAS family of toxins but the report from the Minnesota Department of Health is the oldest surviving complete set of data and narration. In January of 1950, 3M itself conducted tests with PFBA, "perfluorobutyric acid"[31], an early iteration of PFAS. Administered to mice orally, intraperitoneally (directly administered to the abdominal wall), and intravenously the tests showed how rapidly PFBA built up in the bloodstream of mice. Only the data survives (there is no written recap of these tests) but the data clearly shows the worst buildup in blood happened when PFBA was administered to the mice orally, followed by the intravenous injection and finally the intraperitoneal dosing. All three methods produced dangerous levels with the oral dose remaining high even one week after the initial dosage. This, therefore, is the first laboratory evidence of PFAS' trait of bio-accumulation. The ultimate fate of those mice is unknown other than they became highly toxic with extraordinary amounts of PFAS chemicals in their blood. This test, conducted by 3M in 1950, proved PFAS-type substances' toxicity. 3M knew forty-eight years before it admitted it publicly, that PFAS was toxic to mammals. As attorney Rob Bilott says of his class action lawsuit against DuPont, "(i)f I had a case, it was going to be all about toxicity"[32].

---

[31] Exhibit 1009, State of Minnesota v. 3M Co., Court File No. 27
[32] Robert Bilott, *Exposure: Poisoned Water, Corporate Greed and One Lawyer's Twenty-Year Battle Against DuPont* (New York, Atria Books, 2019) pg. 50

On the list of "cute and cuddly" animals, rats would probably occupy a spot somewhere toward the bottom. A few people love them and keep them as pets and companions, most people loath them. A member of that large group of creatures we consider vermin, rats seem to occupy a special place of disgust among most humans. There is the association with the Plague of the Middle Ages that killed millions globally. In movies, Indiana Jones faced hundreds in the sewers beneath Venice and other evil rats like Ben and Willard have been the focus of sinister film deeds. Ancient creatures they have been detested in particular by city dwellers as long as there have been cities. Perhaps it is this less than ideal relationship the rat has with mankind that, when it comes to research, rats are second only to mice as test subjects for a wide variety of products, conditions and processes.

Between 1976 – a year after PFAS was found in a wide sampling of human blood – and 1998 when 3M's role in the worldwide fluorocarbon disaster was discovered, no less than eighteen tests were conducted with rats as the subjects. Out of all those studies and all those rats who were always "sacrificed" at the end of the test, none drew more attention than a test conducted using a PFOA variation called FM-3422. Differing from other tests, this was a teratology study meaning 3M wanted to know if FM-3422/PFOA caused congenital abnormalities: birth defects. The rats used in this study were all female and all pregnant when they were dosed with varying amounts of PFOA. The year was 1980.

The tests, mercifully, lasted only 20 days. The mother rats were, for the most part, unharmed but the fetuses almost universally had "large lens clefts, dark streaks running one-half to three-quarters of the way through the lens or disorganized lens fibers"[33]. In other words, the fetuses were damaged, and the damage centered around the eyes. As mentioned previously, the mother rats were "sacrificed" by "cervical dislocation" a common method of animal euthanasia. It refers to a technique used in physical euthanasia of small animals by "applying pressure to the neck and dislocating the spinal column from the skull or brain"[34].

This test, in particular, would prove to be among the most detrimental to DuPont and 3M. It proved for the first time a direct link between PFAS, pregnant mothers, and their fetuses. After analyzing the results, both companies – 3M and DuPont – removed women of childbearing age from any PFAS production areas. In an internal 3M memorandum dated April 17, 1981, the presiding Medical Director for the Company recommended the following:

> *Females of childbearing potential shall be excluded from working on the production of fluorochemical surfactants and alcohols.*[37]

Fully eighteen years after that, as part of the class action lawsuit in West Virginia, the lawyers who filed the suit conducted depositions as part of the

---

[33] Exhibit 1252, State of Minnesota v. 3M Co., Court File No. 27-CV-10-28862

[34] https://www.google.com/search?ei=Rjn-XdKHJcWwtAa8anIBg&q=oervical+dislocation&oq=oervical+dislocation&gs_l=psyab.3..0i13l10.96131.96131..98454...0.3..0.80.80.1......0....2j1..gws-wiz.......0i71.cr6GFrBmUoA&ved=0ahUKEwjSwtarhsfmAhVFGM0KHbx8CmkQ4dUDCAs&uact=5 [37] Exhibit 1254, State of Minnesota v. 3M Co., Court File No. 27

discovery process. In one of those depositions a lawyer for the litigants was questioning the then Corporate Medical Director from DuPont. This was the exchange regarding that particular test – the PFOA test conducted for 3M on pregnant rats and its results:

> **Attorney**: You did see there was a substantial risk to, ah, women at the DuPont plant who are exposed to C-8 (PFOA) enough to remove them from over-exposure, correct?
>
> **Medical Director**: No, there was no potential risk to the women based upon the 3M study. There was a potential risk to the fetuses.[35]

Tracing the chronology of this teratology test on rats and is aftermath begins to show the lengths both companies went to in avoiding public disclosure of the health risks associated with PFAS. The test was conducted in 1980. After reviewing the results of the test both 3M and DuPont removed women of childbearing age from production locations in 1981 which indicates fairly quick action on the part of both companies. Later in 1981, DuPont surveyed its own employees:

> *"DuPont collected blood data and reviewed records for seven Teflon employees who had recently given birth. All had elevated PFOA levels in their blood. Two of the seven babies had defects noted at birth. Both of those were eye defects. (T)he expected rate of birth defects involving the eyes is two for every one thousand live births (in the general population). DuPont had now found two in seven births."[36]*

---

[35] *The Devil We Know,* Stephanie Soechtig, Jeremy Seifert - Directors, Joshua Kunau – Producer, 2018, YouTube.

[36] Robert Bilott, *Exposure: Poisoned Water, Corporate Greed*

Nearly twenty years after 3M knew of the adverse effects PFOA has on rat fetuses (1980 test and the 1999 deposition), the truth had finally emerged beyond the veil of secrecy held firm by 3M and DuPont. The collusion and conspiracy to hide the truth the test results had exposed was beginning to unravel. Both companies knew in 1980 that PFAS caused birth defects in test animals and, as had become standard PFAS practice, failed to report the results to anyone but each other. Removing women of childbearing age from their own production facilities was probably done more as a tactic to avoid litigation but that strategy would, eighteen long years later, prove to be a costly mistake.

CONTENT WARNING: What follows is, once again, based on court documents from the suit filed by the State of Minnesota against 3M Company in December 2010. The remainder of this chapter deals with multiple tests conducted multiple times on monkeys. What happened during those tests is disturbing; what's even more disturbing is that, dissatisfied with the results of an initial test, 3M requested further tests on monkeys be performed. If you have the slightest issue with the following phrase, you should probably skip forward to the next section: "The study was terminated after 20 days because of the early deaths of the monkeys in all treatment groups".[37] The remainder of this chapter will address how all of those monkeys died for the sake of testing a substance 3M already knew to be toxic: PFAS.

---

and One Lawyer's Twenty-Year Battle Against DuPont (New York, Atria Books, 2019) pg. 177

[37] Exhibit 1009, State of Minnesota v. 3M Co., Court File No. 27

Rhesus monkeys – specifically, Rhesus macaque – are adaptive primates. They have the greatest range of geographic diversity of any non-human primate. They can be found on the plains, in forests and even in the mountains; rural, suburban and urban. Some troops – that's what you call a group of Rhesus macaque, a troop – adapt well to the presence of humans so they are comfortable living along the fringes of many major cities. Perhaps it is that trusting nature, that lack of fear of humans that led them to first be captured and used as test animals. That trusting nature and their anatomical closeness to human beings has meant some Rhesus monkeys have died horrible deaths in labs around the world. PETA estimates that at any given time approximately 106,000 monkeys are being used in laboratory tests in the United States alone. We just don't know how many monkeys died as a result of PFAS testing. Some of those that died because of PFAS testing were a part of tests where the results were never disclosed to the general public. Those trusting Rhesus monkeys, therefore, died in vain at the hands of humans. There should be a reason behind sacrificing monkeys as a part of testing: to benefit human beings. The tests they fell prey to were kept secret by 3M for nearly twenty years. Those test results, had they been shared, might have been used to research a method of dealing with the toxicity of PFAS. These Rhesus were not afraid of humans and in the end gave their lives for humans; gave their lives without recognition from humans. The results were never acted upon. The only known fact derived from these tests was how much PFAS it took to kill a Rhesus macaque. 3M filed that research away, releasing it only under court order after being sued by the State of Minnesota. Those monkeys' gruesome deaths were useless because 3M kept the details of those deaths hidden away until the light of justice brought them into the open.

There was a third test of monkeys which commenced in 1998 the results of which were not published, for some reason, until 2001. Given the timeline of proliferation, this late date meant that it was already too late to put a stop to PFAS proliferation. That being said, it is much more widely accepted that these 2001 findings were more transparent unlike the other two that will be highlighted here.

Not far from Kalamazoo, Michigan is the town of Mattawan, home of International Research and Development Corporation. Since 1962, IRDC has conducted tests for drug, chemical and agricultural customers or as it is called in the sanitized version of its day-to-day work "animal safety evaluation studies". IRDC uses animals to test various substances and it was there, in 1978, that they used sixteen Rhesus macaque in PFAS toxicity tests for 3M. Actually, they conducted two separate studies but more on that in a moment.

The first study was "initiated on February 16, 1978. Terminal sacrifices were conducted on May 17, 1978."[38] Yes, you read that correctly. In what can only be assumed to be the language of animal testing labs, "terminal sacrifices" translates to all sixteen monkeys who participated in this PFAS test for 3M were killed either by PFAS or at the hands of lab personnel. They

---

[38] Exhibit 1191, State of Minnesota v. 3M Co., Court File No. 27-CV-10-28862

were killed regardless of the test results. The control group, who were administered nothing stronger than distilled water during this test, became "terminal sacrifices." But there are gaps in this story to be filled in first.

This first test was conducted using the 3M product "Fluorad® Fluorochemical Surfactant FC-95"[39] in varying doses from 0.5 milligrams per kilogram of each macaque's body weight per day up to 4.5 mg/kg/day. The test ran for ninety days except for those four monkeys in the 4.5 mg. dose group who "died or were sacrificed *in extremis* between week 5 and 7 of the study."[40] In extremis defined as "near death".[41] It will be enough to say that the monkeys in this group died of PFAS poisoning little more than halfway through the 90 day test. Reading the details of how they died and the inhumane journey from the start of the test to their horrific deaths is not something that needs to be shared. What should be the legacy of those innocent creatures is they were used to test the effects of PFAS on humanlike beings and the findings of those tests, findings that could have helped mankind deal with PFAS proliferation and its side-effects on humans were kept callously hidden for two decades by 3M.

The only group that didn't show detrimental effects as a result of PFAS were, obviously, the control group for they were dosed with nothing stronger than distilled water.

---

[39] Exhibit 1191, State of Minnesota v. 3M Co., Court File No. 27-CV-10-28862
[40] Exhibit 1191, State of Minnesota v. 3M Co., Court File No. 27-CV-10-28862
[41] https://www.dictionary.com/browse/in-extremis?s=t

Remember, though, those four unaffected perfectly healthy Rhesus monkeys were also included, for whatever reason, in the "terminal sacrifices" at the end of the test. They killed four monkeys that hadn't even been injected with PFAS for reasons known only to them.

Almost immediately, however, another test began again using Rhesus macaque, five sets of four monkeys – two male, two female per set – with one control group; twenty macaque in total. This time the doses were increased, increased dramatically. The first tests were doses of 0.5 milligrams per kilogram of each monkey's body weight per day, 1.5 mg/kg/day and 4.5 mg/kg/day. There is no supporting documentation so the reason for the massive increases in the doses per day for the second test remain a mystery. Those doses were increased to a staggering 10 mg/kg/day, 30 mg/kg/day, 100 mg/kg/day and an almost incomprehensible 300 milligrams of PFAS solution per kilogram of monkey body weight per day! If the monkeys in the first test receiving 4.5 milligrams per kilogram of body weight per day all died in less than fifty days, why would any right-thinking person believe administering a minimum dose more than twice as much as that would have a different result? The difference in the maximum dose in test one to the maximum dose in test two was 67 times greater. It is amazing to think that those monkeys lasted as long as they did with dosage levels that high.

International Research and Development Corporation is an independent contractor that conducts tests under guidelines established by the customer who, in this case, was 3M. It was 3M who determined the massive and sickening increases in doses. It was 3M who somehow thought the results might be different. It was 3M who signed off on the death warrant for those twenty Rhesus macaque in the second test. And like the pattern established after the first test, shoved the results in a drawer and prevented what might have been human life-saving research from happening.

During those two tests, thirty-six monkeys died in vain because 3M failed to share the results of their deaths with those who might have been able to affect a solution to PFAS poisoning. Those monkeys died at the whim of 3M and their deaths were rendered meaningless by 3M as well. Killing animals to advance science is one thing; killing animals without utilizing test results is inhumane and beyond shameful.

There is no reasonable way 3M can defend its actions when it comes to these macaque tests. Deplorable experiments which claim the lives of monkeys were horrific in and of themselves but to never use the results as a starting point into research that would benefit humans is contemptable.

CHAPTER FIVE

The Story of a Dumpsite – "A Definite Hazard Exists in Regard to Ground Water Pollution"[42]

"Hope is like one of those orchids that grows around toxic waste: lovely in itself - and an assertion, if you like, of indefatigable good - but a sure sign that something nasty lies underneath." - Rachel Cusk

Any manufacturing plant creates waste in both solid and liquid form. Solids can be as simple as wooden pallets and steel drums used to ship and store ingredients or the finished goods themselves. Then there's the liquids. Today, through the Toxic Substances Control Act (TSCA), the disposal of any liquid is a highly regulated process, especially for ones with any degree of toxicity. Harkening back to the early 1960's, however, finds us in a far different world; a world saturated in dangerous and harmful products carelessly, yet not illegally, disposed of. From that perspective, it was a far more dangerous time with few laws to protect the environment let alone human beings.

Shortly after full scale operations began in Cottage Grove, vast amounts of waste, both liquid and solid, were being generated. As

---

[42] Exhibit 1025, State of Minnesota v. 3M Co., Court File No. 27-CV-10-28862, pg. 2

mentioned in Chapter Two, disposal, from the start, was conducted on-site in Cottage Grove. Liquids at the time were run through a soon-to-be-ineffective waste treatment facility on the grounds. Solid waste was burnt or buried right there at the plant.

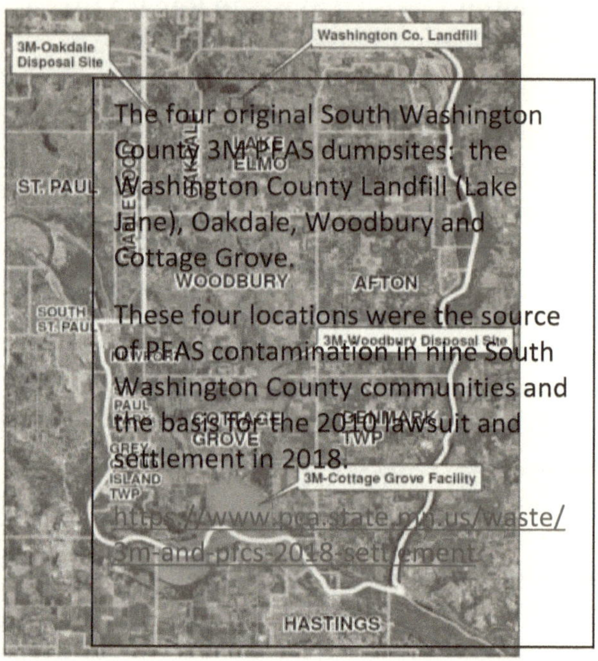

The four original South Washington County 3M PFAS dumpsites: the Washington County Landfill (Lake Jane), Oakdale, Woodbury and Cottage Grove.

These four locations were the source of PFAS contamination in nine South Washington County communities and the basis for the 2010 lawsuit and settlement in 2018.

https://www.pca.state.mn.us/waste/3m-and-pfcs-2018-settlement/

    The second 3M dumpsite mentioned previously was in Oakdale which would, because of its cocktail of toxic ingredients in addition to PFAS, eventually qualify as for cleanup under CERCLA (Comprehensive Environmental Response, Compensation, and Liability Act) known also as Superfund. There were two other dumpsites in South Washington County: the Washington County landfill which is sometimes referred to as the Lake Jane site, and Woodbury.

The Lake Jane site actually lies in close proximity to Lake Elmo, a 206-acre body of water so contaminated by seepage from the Lake Jane landfill that eating fish caught in the lake is inadvisable. In the land of 10,000 lakes there are thirty-three additional bodies of water in the Twin Cities metropolitan area that are so PFAS tainted that eating the fish from them is not recommended for women who are or may become pregnant, and children under age 15. There are a lot of metropolitan areas in this country that don't even have thirty-three lakes yet that is the total number in Minneapolis and St. Paul where consuming fish caught there is not recommended for pregnant woman and children. Among the thirty-three are prominent, highly utilized lakes like Lake Calhoun (recently renamed Bde Maka Ska), Lake Como, the previously mentioned Lake Elmo, Lake Harriet, Lake Johanna, Keller

Lake, Lake of the Isles and Lake Phalen. In addition to lakes, both the state of Minnesota and the state of Wisconsin have issued PFAS-based advisories on consumption of fish caught in the Mississippi River ranging from no more than once per week to never.

Those four sites – the plant site in Cottage Grove, Oakdale, Lake Jane/Washington County Landfill and Woodbury – were the originators of the contamination that was the basis for the 2010 lawsuit filed against 3M by the state of Minnesota on behalf of the Minnesota Pollution Control Agency (MPCA) and the Minnesota DNR, Department of Natural Resources. They were the root cause of the tainted water found in the nine communities named in the suit.

In the treasure trove of documents uncovered by that litigation the most well documented of the four 3M dumpsites is Woodbury. From July 13, 1960 when the initial site recommendation was filed by the 3M Geology Department until its closure on April 29, 1966, the toxic tale of the 3M Woodbury dumpsite is fully documented. In many ways, Woodbury was typical of the other sites so rather than delving into similar dumpsite histories of all four locations, we will look at the tale of Woodbury's from start to finish. As will become clear, this is a tale without a happy ending.

A suburb east of the Twin Cities of Minneapolis and St. Paul, Woodbury became a PFAS dumpsite in 1960.
https://www.durhamexecutivegroup.com/woodbury-mn-home-values/

        "Bucolic" is the word that appears in
many descriptions of Woodbury. It is one of
those communities urbanites point to with
derision since it is the very embodiment of
"suburban" living: single-family homes,
manicured lawns, miniscule crime rate, a highly
regarded school system, multiple golf courses
and multiple shopping venues.   The city "violent
crime rate for Woodbury in 2016 was lower
than the national violent crime rate average by
83.22% and the city property crime rate in
Woodbury was lower than the national property
crime rate average by 20.85%."[43] In other
words, Woodbury is a quiet and safe place to
live. In 2018, Woodbury was named by *Money*
magazine as one of the "10 Best Places to Live in
the United States"[44]. It also placed second on
the "Best Cities to Raise a Family for 2020" list
compiled by homesnacks.com[45]. In 1960,
Woodbury, then called Woodbury Township
since it had not yet attained city status, was a
second ring community of 3,014[46] people and
had a livestock population that was probably
twice that size. Sixty years later, no longer
sleepy, no longer rural, Woodbury's population
has skyrocketed to an estimated 71,306[47]. In

---

[43] https://www.cityrating.com/crime-
statistics/minnesota/woodbury.html
[44] https://money.com/collection/2018-best-places-to-
live/5361448/woodbury-minnesota-2-2/
[45] https://www.homesnacks.net/best-cities-for-families-in-
minnesota-1211110/
[46] https://population.us/mn/woodbury/
[47] http://worldpopulationreview.com/us-cities/woodbury-mn-

many ways, both light and dark, Woodbury is
the best-kept secret in Minnesota.

Woodbury is a contemporary success story
with a haunting past that, quite literally, lurks just
beneath the surface. In the vast accumulation of
discovery documents from the 2010 Minnesota lawsuit,
the first mention of Woodbury appears in an internal
3M memo dated July 13, 1960 and titled "Geology
Dept. Report #60-10"[48]. It is, as the title implies, a
report by the internal 3M Company Geology
Department on a piece of land in then Woodbury
Township that was, at the time, owned by a company
called Terminal Warehouse. Terminal Warehouse's
business was industrial and commercial waste disposal
both solid and liquid.

Right off the bat something seems amiss.
Mentioned in this geology report is the fact that
ownership of the site would continue to be held in the
name of the present owner "to minimize publicity"[49], a
rather curious statement for a geology report. It also
gives the appearance 3M did not want its name
associated with this particular dumpsite. Located in the
southeast corner of Woodbury Township along its
border with Cottage Grove, 3M was considering the
site as the latest in a series of landfills it would use as
disposal locations for waste from its recently renamed
Chemolite plant in Cottage Grove. Disposal continued
on-site at the plant and also at Oakdale, but the
Oakdale location was nearing its limits; an additional

---

population/

[48] Exhibit 1025, State of Minnesota v. 3M Co., Court File No.
27-CV-10-28862
[49] -CV-10-28862

site, a third site was needed after little more than ten years of manufacturing PFOA in Cottage Grove.

The July 13th report mixes geological data with general information concluding with "(f)rom our preliminary survey we feel that this area will make a suitable dumping site."[50] Before reaching that conclusion, however, the report makes some interesting statements.

"The wet waste material will be dumped in long trenches and allowed to seep into the ground and as soon as one area becomes saturated, it will be covered over and another trench dug."[51]

That statement mentioning seeping wet waste materials comes immediately after those waste materials are identified as "wet acid and phenol waste"[52][53]. Not mentioned is the residue from the manufacture of PFOA. How far would that toxic concoction of acids and PFOA have to "seep" before it reached the groundwater? The report itself claims:

"It appears that the groundwater elevation in this area is approximately 860 feet above sea level and our principal dumping area will be at about 900 foot(sp)."[56]

---

[50] Exhibit 1025, State of Minnesota v. 3M Co., Court File No. 27-CV-10-28862, pg. 3

[51] Exhibit 1025, State of Minnesota v. 3M Co., Court File No. 2728862, pg. 1

[52] Exhibit 1025, State of Minnesota v. 3M Co., Court File No. 2728862, pg. 1

[53] , pg. 2

The answer then to the question of the distance between the waste and the water? A mere forty feet. "When waste seepage reaches the water table certain solids by that time may be rendered harmless while others will remain to pollute wells"[57] sounds like 3M knew with a degree of certainty that, sooner or later, the waste being dumped on this site would end up in drinking water.

As quoted briefly earlier in this work, the report goes on to say:

"The zoning problem alone merits a good deal of attention. Low level land usage such as dumps, and gravel pits generally bring condemnation from nearby residents and population centers. This cannot be over-emphasized after past experience with our eastern zoning problems."[58]

Exactly what those "eastern zoning problems" were is not mentioned in this report or any other document in the collection but its presence suggests this was not the first time 3M had issues with the zoning of dumpsites. Undeterred, the very next sentence demands "(i)t is imperative that we have absolute freedom to use the land for dumping before purchase and use"[59] which strikes the reader as "damn the rules and zoning regulations, we're going to use this dump as we see fit". And this was before dumping had even begun. It also seems to be saying 3M intended to start dumping immediately even before transfer of ownership from Terminal Warehouse.

investigated"[54] making it very clear that after about eighteen months of operation by 3M, the dumpsite already had issues.

After drilling core samples from three different locations at the site, the 3M Water and Sanitary Engineering Department found "(t)he ground water level is about 75' below ground surface"[55] and through those test borings it was found that "acetone was penetrated to the level of 75'"[56] (groundwater level) at the first bore location and "ethanol, heptane, acetone, ethyl acetate, and methyl ethyl ketone were found at the elevation of 64' below ground"[63], eleven feet from the ground water in boring number two. This, as the report notes, was "within about one year of operation"[64]. What those tests didn't find was PFOA since they weren't specifically looking for it and you have to run specific tests to find it. Also, the testing methods of the time precluded testing for PFAS levels. It was there, mixed in with the acids and acetones. Since this was liquid waste from a plant producing PFAS there is no possible way liquid waste from Cottage Grove was ever free of PFAS; it is an absolute, sure-fire certainty.

Based on their findings, the 3M Water and Sanitary Engineering Department made three sets of recommendations. Their first conclusion simply states: "Laboratory data substantiates the fact that waste chemicals are reaching the ground water level at the Woodbury Dump Site"[65]. Recommendation number one goes on to say that "all future waste dump pits be lined"[57]

---

[54] Exhibit 2621, State of Minnesota v. 3M Co., Court File No. 27-CV-10-28862, pg. 1
[55] Exhibit 2621, State of Minnesota v. 3M Co., Court File No. 27-CV-10-28862, pg. 1
[56] Exhibit 2621, State of Minnesota v. 3M Co., Court File No. 27-CV-10-28862, pg. 1
[57] Exhibit 2621, State of Minnesota v. 3M

which openly admits none of the current pits are lined. In other words, from its inception until it was closed, the 3M Woodbury dump site had pits that were unlined which means waste, including waste from PFOA production, was just dumped on top of the ground and allowed to soak uninhibited through the layers of dirt and rock straight through to the ground water and aquafers. Akin to closing the barn door after the horse has escaped, it is complete folly to line an unlined pit once the chemicals in that pit have already seeped into the ground.

The first recommendation tells us 3M had proven waste chemicals had reached the groundwater and since we have also established the certainty the waste chemicals included PFAS, we have reached the conclusion that on May 16, 1962, 3M knew it had polluted the groundwater in Woodbury with PFAS; one can come to no other conclusion.

The second recommendation starts out this way: "In view of the publicity of the ground water pollution in Newport (a small community on the Mississippi about eight miles northwest of Cottage Grove), which was publicized in the *St. Paul Dispatch* last March, by Commercial Chemicals who are handling our wet scrap"[67] you don't need to read further. This current situation in Woodbury was not the first time 3M had polluted groundwater since there had been another in a dump where 3M chemicals had been disposed of through an outside contractor two months previous in the nearby community of Newport. Learning from the mistakes made two months earlier, this report thinks it would be a good

---

Co., Court File No. 27   28862, pg. 2 [67]
Exhibit 2621, State of Minnesota v. 3M Co.,
Court File No. 27 28862, pg. 2

idea to drill some observation wells to keep track of how fast this site in Woodbury was polluting.

The final recommendation, which is no solution at all, says "(a)fter a final solution of disposal for our wet scrap is found … the acid wet waste remains to be continuously disposed into the ground at the Woodbury site"[58] which, to the untrained eye appears to be saying we'll just keep dumping liquid acid waste and PFAS residue directly into the ground in Woodbury until we just can't do it anymore.

The year 1963 has three documents in the discovery documents database. The first, from July 10th of that year, contains this statement: "Because of the decision not to line the trenches, a definite hazard exists in regard to ground water pollution"[59], part of which was used for the title of this chapter. Deconstructing that sentence we find two damning statements. The first being the admission that 3M knew the pits, which were supposed to be lined to prevent seepage, were not. Time and again they skirt the issue of the pits and if they were or were no lined but here, in 1963, they admit they knowingly made the decision not to line those pits. The second part of that sentence is also an admission this time 3M is fessing up to knowing they have polluted the groundwater in Woodbury. 3M knowingly didn't line the pits which caused, in their own words. "a definite hazard" to the groundwater.

---

58 Exhibit 2621, State of Minnesota v. 3M Co., Court File No. 27-CV-10-28862, pg. 2
59 Exhibit 1043, State of Minnesota v. 3M Co., Court File No. 27-CV-10-28862

Something in this memo must have touched a nerve at 3M because a little more than two weeks letter on July 26ᵗʰ a second piece of interoffice correspondence appears recapping a second site visit. The purpose of this visit was for "determining if any improvements have been made in waste disposal"[60]. It includes this passage:

"A discussion with one of the truck drivers brought out the fact that the trench used for flowing wet waste had been lined with bentonite in October 1962. It appears to the writer that this seal is ineffective. The truck driver pointed out the fact that as soon as the waste was dumped it seeped into the ground."[61]

The trench being referred to was one that was lined using the substance – bentonite – 3M had determined would solve the seepage issue. That conversation, then, most likely put an end to any thoughts anyone might have had that "any improvements have been made in waste disposal".

The final document from 1963, dated December 2, 1963 in the database, gives an overview of the present operations as they pertain to waste disposal. Under a section entitled "Unpumpable Residue" is the following passage:

---

[60] Exhibit 1043, State of Minnesota v. 3M Co., Court File No. 27-CV-10-28862
[61] Exhibit 1043, State of Minnesota v. 3M Co., Court File No. 27-CV-10-28862

"Certain waste materials are of such consistency that they cannot be pumped and thus cannot be incinerated. These materials consist of resins such as acrylates and alkyds and rubber like compounds which have settled from the liquid scrap. The unpumpable residue is sorted from the pumpable materials (liquid scrap) and disposed in lined pits at the Woodbury sites."[62]

This passage follows others from mere months earlier that say there are no lined pits in Woodbury and those that were lined were "ineffective". This document is highly suspicious. It carries no author's name, who it was prepared for or how it was to be used but in reading it one gets the impression this is in response to some formal complaint from perhaps a higher authority. Was 3M questioned by some governmental agency on how it disposed of its solid and liquid wastes? We are unable to tell since there is no other document in the database referencing this particular piece of information. One can only hope 3M wasn't already in a position of lying to the government.

Nineteen-sixty-six became the seminal year in the life of the 3M Woodbury dumpsite. On April 18th of that year a homeowner adjacent to the site complained of foul odors and a bad taste in his well water. Based on that complaint, 3M tested wells in eighteen adjacent properties and found eleven of the eighteen contaminated to one degree or another. The Woodbury site was permanently closed on April 29, 1966 eleven days after that initial complaint. Moving that quickly — eleven days from the first complaint to completely shutting down the facility — seems extraordinary. Someone at 3M must have sensed, or perhaps already knew, something was horribly wrong

---

[62] Exhibit 1046, State of Minnesota v. 3M Co., Court File No. 27-CV-10-28862, pg. 1

with the Woodbury dumpsite. The volume of documents about the 3M Woodbury dumpsite is all negative. The documents start up raising questions and end with the site being shut down in less than three years. After the closure the database is awash in documents dealing with the closing of the site and covering the toxic tracks that would become the dumpsite's legacy.

Two months after the initial complaint, "an activated carbon filter was installed in the (homeowner's) well"[63]. With 20/20 hindsight, we can now look back over the span of fifty-four years and realize Woodbury homeowners by the hundreds are now purchasing and installing POET (Point of Entry Treatment) carbon filters as a solution to contaminated drinking water. They are implementing the same solution 3M provided all those years ago for that single tainted house. It has now become an option for the remediation of the problem they created back then. The only difference is 3M paid for that filter back in 1966; they're not picking up the tab today.

The complaints and subsequent closing of the 3M Woodbury dumpsite drew unwanted attention, so a public township meeting was called a mere four months after the site was closed. Attendees were told that, upon testing the water at the suspect private well, 3M found "high concentrations of surfactants"[64] (remember: PFAS, among other things, is a surfactant) but attributed their presence to local pollution namely laundry detergent. Then, in the very next sentence in the meeting minutes, admits "3M could not find any trace organic chemicals (PFAS is also an organic chemical) ...with our present laboratory techniques."[65] That is a bit of machination. 3M knew it

---

[63] Exhibit 1058, State of Minnesota v. 3M Co., Court File No. 27-CV-10-28862, pg. 5

[64] Exhibit 2589, State of Minnesota v. 3M Co., Court File No. 27-CV-10-28862, pg. 2

was manufacturing fluorochemicals at Chemolite in Cottage Grove. It stands to reason that the residue from that manufacturing process would almost certainly contain fluorochemicals. One can then conclude that the wet waste from that plant being disposed of at the 3M Woodbury dumpsite would have to include fluorochemicals; you wouldn't need to go beyond "present laboratory techniques" to determine that. All you would need is good old-fashioned common sense and logic.

At the same meeting, government culpability is raised for the first time:

> *The local state representative "wanted to know why the Township Board and the Minnesota Health Department and Water Pollution Control Commission originally allowed a dump of this type to be established in this area. Since the Health Department had not issued 3M an operation permit for the dump, he also wanted to know why 3M was allowed to continually use the site."[66]*

"Since the Health Department had not issued 3M an operation permit for the dump" seems to be an overwhelming condemnation that draws a resultant plethora of questions. Questions like Why? Why hadn't the Health Department issued a permit? Did the Health Department refuse to issue an operation permit? Since they had operated other dumpsites before, why didn't 3M

---

[65] Exhibit 2589, State of Minnesota v. 3M Co., Court File No. 27-CV-10-28862, pg. 2

[66] Exhibit 2589, State of Minnesota v. 3M Co., Court File No. 27-CV-10-28862, pg. 3

question not being issued an operation permit? Better yet, since they would have been issued operation permits for other sites before why would 3M go ahead and continue dumping in Woodbury without one? With that statement it becomes clear that either one of two things happened: 3M, without an operation permit, illegally dumped PFAS contaminated waste in Woodbury flagrantly breaking the law; or the Minnesota Health Department was nothing shy of lax in its obligations and failed to issue a permit. Or maybe it was a combination of both. Regardless of which it is, we now know that the 3M Woodbury site was unregulated; 3M dumped product at will knowing full well they had not been granted governmental permission to do so. Skip ahead to present day: this lack of regulatory involvement has become part of 3M spin. They tell employees and anyone else that will listen that what they did back then wasn't illegal an excuse that seems rooted in this lack of permitting by the Minnesota Health Department back in the

1960's. This is a sterling example of what former Minnesota Attorney General Lori Swanson meant when she said, "3M had so many opportunities to do the right thing, but they never did." They not only have a financial obligation to their community and their stockholders, 3M also has a moral obligation to be a good citizen and a good neighbor. This incident smacks of "getting away with something" and in this case what they got away with was the dumping of a pollutant that would contaminate the world.

Late in 1966, a document quantifies the extent of the abuse of the land, water and air around the 3M Woodbury site. "Approximately 6.0 million gallons of wet scrap was disposed at the Woodbury site"[67] and

---

[67] Exhibit 1058, State of Minnesota v. 3M Co., Court File No. 27-CV-10-28862, pg. 6

"(a)pproximately 50% of this material was disposed during the period from 1960 to 1963 in unlined pits".[68] The last half of that sentence is one of the most damning in the collection. A sentence within a 3M internal document that says they dumped directly onto the ground in unlined pits about three million gallons of highly suspicious wet manufacturing scrap from a plant that produced, among other things, tons and tons of PFAS products. Three million gallons would fill up 75,000 bathtubs.[79] According to the EPA, the average household uses about 400 gallons of water each day so three million gallons would be more than a twenty-year supply.[69] Lastly, it would take 22,721,893.5 half-liter bottles of water to account for three million gallons. But we are talking about industrial waste that was just dumped into unlined pits over a three-year span. The totality of that deed is seemingly incomprehensible.

The year 1966 concludes with one last internal 3M document titled "WATER POLLUTION PROBLEM IN WOODBURY".[70] This piece reviews the situation and for the first time reveals the fact that "(p)resent data indicates contamination not yet widespread but apparently moving in a way that could affect neighboring farm water supplies and ultimately the water supply for St. Paul Park, Newport and Thompson Grove."[71] That water supply mentioned is "the Jordan geologic formation is the principle aquifer in

---

[68] Exhibit 1058, State of Minnesota v. 3M Co., Court File No. 27-CV-10-28862, pg. 6 [79]https://www.google.com/

[69] https://www.google.com/

[70] Exhibit 2639, State of Minnesota v. 3M Co., Court File No. 27-CV-10-28862

[71] Exhibit 2639, State of Minnesota v. 3M Co., Court File No. 27-CV-10-28862

the Minneapolis-St. Paul metropolitan area and is widely utilized for domestic, industrial and municipal water supplies".[72] As we would later learn, that "could" turned into "did" as the Jordan aquafer was infiltrated by PFAS. The memo also raises concerns about the clean-up solutions and how those remedies would come "at a very substantial cost to 3M."[73] The memo concludes: "Obviously, there must be no discussion outside the company except by those having directly assigned responsibility in this area."[74] This was 1966 and the cover-up had begun in earnest.

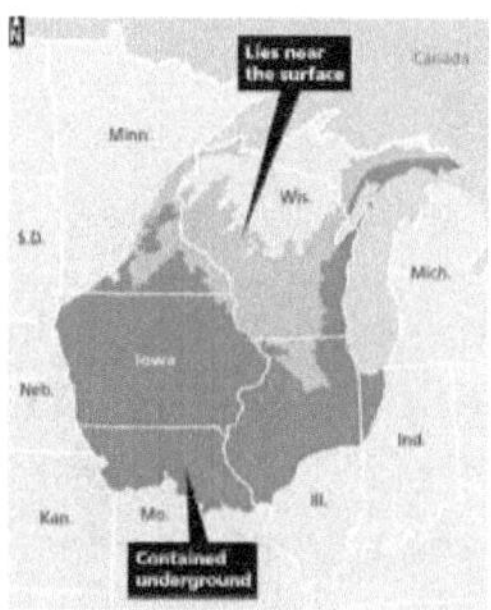

Jordan aquifer

The Jordan aquifer, officially the Cambrian-Ordovician aquifer system, touches seven states and covers most of Iowa, where it is mostly underground. Experts say the aquifer needs decades to recharge. In many places, including parts of Iowa, the water is being drawn from the aquifer faster than it can be recharged.

Source: U.S. Geological Survey

At this point, let's put the Jordan Aquifer in perspective. First of all, it is massive running from the Upper Peninsula of Michigan through Minnesota, Wisconsin, Illinois, Indiana on through almost all of Iowa and a portion of Missouri. The northern portion of the

---

[72] Exhibit 2660, State of Minnesota v. 3M Co., Court File No. 27-CV-10-28862, pg. 12

[73] Exhibit 2639, State of Minnesota v. 3M Co., Court File No. 27-CV-10-28862

[74] Exhibit 2639, State of Minnesota v. 3M Co., Court File No. 27-CV-10-28862

Jordan is very close to the surface as is Woodbury, Oakdale, Cottage Grove and Lake Jane. The rest is deep underground and it is not fully known whether that portion is polluted with PFAS. It is understandable why 3M would not want to admit that it had polluted the Jordan Aquifer with PFAS. As previously mentioned, the groundwater supply in Woodbury is the Jordan where it is extremely close to the surface. The 3M Geology Department report of 1960 pointed that out. When PFAS entered the groundwater in Woodbury – and probably Cottage Grove, Oakdale and Lake Elmo as well – it was a direct pipeline into the Jordan Aquifer.

As 1966 turned into 1967, a report from an independent hydrologist appeared assessing the Woodbury situation. Looking back fifty-three years we are struck with an eerie coincidence. In a section on Groundwater Quality while referencing one of the Woodbury test wells the reports says "(t)he water also created a noticeable foam which was not persistent and rapidly disappeared"[75]. Fast-forward to January 14, 2020 and an article in the Minneapolis *Star Tribune* which said, in part, "Elevated levels of the per- and polyfluoroalkyl substances, called PFAS, turned up in naturally-occurring foam in Raleigh Creek in Oakdale in Washington County, and in Battle Creek in eastern St. Paul in Ramsey County."[76] Foaming water in 1967 and foaming water in 2020 seem more than a little coincidental.

Amongst the conclusions in the 1967 hydrologist's report is "(t)he chemically contaminated ground- water is objectionable and may be a health

---

[75] Exhibit 2660, State of Minnesota v. 3M Co., Court File No. 27-CV-10-28862, pg. 9

[76] Jennifer Bjorhus, "PFAS 'forever chemicals' found in foam in two east metro creeks", Minneapolis Star Tribune, January 14, 2020

hazard."[88] This report also suggested removing the chemically laced dirt and burning it on site in Woodbury.

On March 23, 1967, 3M met with the Chairman of the Minnesota Water Pollution Control Commission and a geologist from the state Department of Health at a clandestine location miles away from the headquarters of any of those entities. This meeting smacks of being covert. Mentioned early on in the written report by 3M is a passage in reference to the WPCC which the writer says "they, too, wished to avoid any undue publicity."[89] Topics included the plan to have "representatives from 3M meet informally with the entire commission for lunch"[90] to discuss the company's desire to discharge the tainted water from the Woodbury site directly into the Mississippi River. Collusion continues as the memo mentions that, at the approval meeting following the free lunch "it is unlikely that any reporters will be present".[77] The approval process and vote "will not be listed on the official agenda of the coming Commission meeting"[92] meaning there would be no public record of the vote ever taking place. The vote did take place in secret and 3M, after providing the proverbial "free lunch" for everyone who took part in that vote, got their way. It was obvious in reading this memo that the 1967 Minnesota Water Pollution Control Commission was already in the pocket of 3M.

---

[77] Exhibit 2640, State of Minnesota v. 3M Co., Court File No. 27-CV-10-28862, pg. 1

The previously mentioned plan to remove the contaminated dirt and burn it onsite in Woodbury was tested and approved. Although no final numbers are in the public record, it was estimated "(t)he burning will take about ten weeks, 24 hours each day, seven days each week."[93] Seventy days of continuous burning of PFAS riddled soil that was approved by both Woodbury and Cottage Grove even after 3M admitted the process "could possibly be harmful".[94] Additional governmental agencies turning a blind eye to 3M's dangerous ideas and solutions to an already hazardous situation. The result of that burning was seventy days of PFAS-laced fumes and smoke being released into the air above Woodbury and being carried by prevailing winds to points far beyond South Washington County, Minnesota.

Early the following year, 1968, someone at 3M was being interviewed probably by a member of the media. As is sometimes the case with interviews, the interviewee needed to be briefed beforehand on what they should and shouldn't say. When you have been determined to be the source of massive amounts of ground and water contamination seeping from a company-owned dumpsite, that briefing sheet can make for interesting reading fifty-two years down the road. "Topics to avoid: Specific reference to how much waste was buried".[78] That's just the beginning of the deceit.

---

[78] Exhibit 1068, State of Minnesota v. 3M Co., Court File No. 27-CV-10-28862, pg. 1

Under the question "what is the extent of the problem"[79] the points to avoid include:

- "Any admission the Jordan water table may be contaminated"[97] which, of course flies in the face of the facts spelled out in the independent hydrologist's report from one year earlier.

- "Any reference to the acid pits"[98] another fabrication or misdirection since the acid pits, by that time, had been around for nearly eight years.

Under the points to stress, the interviewee was told to lie:

- "Only 1 well has been affected"[99] which is a lie since 11 of 18 wells tested back in 1966 were proven to be contaminated.

- "There is no threat to any other private homes in the area"[80] which is about as egregious of a lie ever concocted.

This document, number 1068 in the collection of discovery documents, marks the beginning of the subterfuge, deceit and misdirection employed by 3M to mislead the public. Never confess, hide the facts, distort the truth but most of all, don't get caught.

One last stop on the paper trail that is the historical record of the 3M Woodbury dumpsite. In 1980, an unidentified independent organization – perhaps governmental – wrote an extensive history

---

[79] Exhibit 1068, State of Minnesota v. 3M Co., Court File No. 27-CV-10-28862, pg. 3

[80] Exhibit 1068, State of Minnesota v. 3M Co., Court File No. 2728862, pg. 3

of the site and its toxic past. If ever there was a succinct recap of the legacy of the 3M Woodbury dumpsite it was captured in this document:

> *"One of the biggest problems associated with the disposal pits at the Woodbury site was that the Company had no idea as to the exact nature and quantity of wastes dumped into the pits."*[81]

Put another way, 3M, for six years, dumped unknown quantities of unknown wet waste into the environment without forethought or testing. They went ahead and dumped toxic waste anyway, thoughtlessly, carelessly, recklessly, irresponsibly.

---

[81] Exhibit 2662, State of Minnesota v. 3M Co., Court File No. 2728862, pg. 9

3M Woodbury dumpsite today, barren of any substantial growth but full of warnings and hints of its past.

https://www.mprnews.org/story/2007/05/11/contamdeal

The Woodbury dumpsite today is open land surrounded by fencing that is dotted with blue and white signs reading: "No Trespassing -- 3M Property" as if any rightminded, knowledgeable individual would ever want to do that. What is striking about this tract of land is how barren it is, its lack of trees and substantial growth. Weeds and scrub brush are the most abundant plant life. Woodbury is growing to the south of the city – toward the site -- since it has no room to expand in any other direction. The land to the south, the land that was home to the 3M Woodbury dumpsite, is precious since it accounts for nearly all of the remaining undeveloped land in the city. Homes, a megachurch and a dog park are among the new neighbors to the dumpsite. A neighboring church has a well. That well was tested in 2017 and its

PFAS level was 600 ppt[82]. EPA acceptable limit of
PFAS in drinking water is 70 ppt or below.

This church's well is 8.5 times greater than the
federal level and 16.5 times greater than Minnesota's
acceptable level of 36 ppt.

Huge subdivisions of expensive homes are
being built within walking distance of the 3M
Woodbury dumpsite homes which, fortunately, are
being connected to the Woodbury public water
system which is monitored far more frequently
than nearly all other water supplies in the state.
But the owners of those homes are, for the most
part, blissfully unaware of the lurid past of that
open space just down the road or in some cases,
just across the street.

My home is about two miles to the north of
the 3M Woodbury dumpsite and when we bought
it in 2013, no one told us of the evil down the road.
To this day, 3M continues to "avoid any undue
publicity"[83] by creating an alternate reality. One
would like to think that we, as a nation, have made
monumental progress on environmental issues
since 1966. From the creation of the
Environmental Protection Agency in 1972 to 3M's

---

[82] https://www.ewg.org/interactive-maps/2019_pfas_contamination/map/?_ga=2.83688918.1350710445.1579527447-159866873.1575744099
[83] Exhibit 2640, State of Minnesota v. 3M Co., Court File No. 27-CV-10-28862, pg. 1

disclosure to that agency in 1998, the EPA was unaware of the dangers of PFAS. Bill Clinton was President on that May day in '98 and down through the administrations of three subsequent Presidents – Democrat and Republican – little progress has been made of curtailing PFAS proliferation and its harmful effects.

When looking for an entity that has failed miserably in their responsibilities for protecting the American people from the specter of PFAS, look no further than the Environmental Protection Agency (EPA). This is the agency that's mission is to "protect human health and the environment" yet they also made the following statements:
"EPA has the authority to set enforceable Maximum Contaminant Levels (MCLs) for specific chemicals and can require monitoring of public water supplies".[84] "There are currently no MCLs established for PFAS chemicals".[85] Let me translate that from government-speak: the EPA sets no enforceable standards for PFAS in the drinking water supply anywhere in the United States of America.

When it comes to the issue of how PFAS has interfered with EPA's mission to "protect human health and the environment" we know certain irrefutable facts:

- EPA was informed of PFAS toxicity and its presence in global human sera samples by 3M in 1998

---

[84] https://www.epa.gov/pfas/pfas-laws-and-regulations
[85] https://www.epa.gov/pfas/pfas-laws-and-regulations

- PFAS is present at varying levels in groundwater in forty-nine of fifty states or 98% of the country

- Consistent, on-going monitoring of local drinking water supplies is conducted only on a local basis, not by the EPA

- Nine states currently have regulations that impose stricter PFAS standards than what is called for by the EPA

Consider this statement made by EPA on January 7, 2020: "Under President Trump, EPA is continuing to aggressively implement our PFAS Action Plan – the most comprehensive cross-agency plan ever to address an emerging chemical of concern". Emerging? If you compare that statement with fact, EPA believes PFAS has been "emerging" for over twenty years since it was told of its presence in humans as early as 1998.

There is also this EPA statement from a month earlier in December 2019: "EPA released an ANPRM (Advanced Notice of Proposed Rule Making) to consider adding PFAS chemicals to the TRI (Toxics Release Inventory), a publicly available database where manufacturers annually disclose the quantities of certain chemicals they release into the environment, recycle, incinerate or otherwise dispose of". This statement requires a two-part translation: first, EPA is "considering" "proposing" putting PFAS on a list of entities who release PFAS into the environment. Considering proposing. The second part of the translation is they are considering proposing adding PFAS to a list of organizations who have already released PFAS into the environment. And you might ask what good is that?

One last PFAS-related EPA statement: "On December 3, 2019, EPA sent a proposed regulatory determination for PFOA and PFOS to the Office of Management and Budget for interagency review. Once interagency review is complete, EPA will submit the proposal to the federal register for public comment. The proposed regulatory determination is the next step in the national primary drinking water standard setting process under the Safe Drinking Water Act." That OMB interagency review mentioned in this statement typically takes six to nine months. After over twenty years, what's another six to nine months of inaction by EPA.

On May 3, 2019 over 180 countries agreed to ban production and use of PFOA and its related components. The United States – birthplace of PFOA where the governmental agency in charge of the environment knew PFOA was present in human blood samples as early as 1998 – is not one of those signatories.

CHAPTER SIX

AFFF: Polluting the U.S. Military and Beyond

"Our lives begin to end the day we become silent about things that matter." - Martin Luther King, Jr.

When it came time to select the first-ever Secretary of Defense, thirty-third President of the United States Harry Truman chose the contentious and prickly James Forrestal. Forrestal had forsaken a successful Wall Street career to assist in the nation's effort during World War II. First as Under Secretary of the Navy then the head man, he spent seven years guiding the nation's fleet. He butted heads with Truman by being pro-Arab but the two were aligned in their mutual distrust of the Soviet Union.

In 1952 the United States Navy laid the keel for the world's first supercarrier aircraft carrier. When the new carrier was commissioned on October 1, 1955 it would be named after the first-ever Secretary of Defense: the *USS Forrestal* with the motto "First in Defense".

The U.S. Navy aircraft carrier USS *Forrestal* CVA -59

U.S. Navy photo

The *Forrestal* would quite literally sail to all corners of the globe. On 29 July 29, 1967 the *Forrestal* was in the Gulf of Tonkin stationed over the coast of Viet Nam. A rocket on an F-4 Phantom Navy jet fighter misfired, striking another aircraft. The rocket's impact dislodged and ruptured the other plane's 400-gallon external fuel tank. Fuel from the leaking tank caught fire, causing a raging inferno that burned for hours. When the fire was finally extinguished, 134 were dead, 161 injured or burned, and 21 aircraft destroyed. When the costs of the tragedy were added up, the fire on the *Forrestal* would cost the Navy $72 million. As a historical footnote, on the flight deck that day was a Navy Lieutenant Commander named John McCain.

Four years earlier, in 1963, the Navy Department had solicited the help of 3M in creating a revolutionary firefighting chemical. Not knowing the tragedy that lay ahead, the Navy wanted to, more specifically, fight fuel fires more effectively. The Navy and 3M worked together on the project but after the Forrestal tragedy the need became apparent and this new, unique fire-fighting product was adapted across all United States military entities as the standard for putting out fuel fires. AFFF was put into everyday use because of the Forrestal tragedy. Make no mistake, AFFF – **A**queous **F**ilm **F**orming **F**oam – when used as a firefighting agent has, in its lifetime, saved thousands of lives and billions of dollars in property. Those savings came with a cost. Any Grade B fire extinguisher whether it be the one under your kitchen sink, in the hallway of your children's school or by the hundreds of gallons in fire trucks at nearly every airport in the United States and around the globe contains PFAS; large quantities of PFAS. It has been estimated that the PFAS contained in AFFF has been the largest single contributor to PFAS proliferation.

AFFF was the best thing ever discovered for putting out flammable liquid fires; flammable liquids like gasoline and jet fuel, grease fires and lighter fluid. An AFFF fire extinguisher mixes a recipe of air, water and a foam concentrate (an ingredient of which is PFOS) and expels it, through a nozzle, onto the fire. Since PFOS is highly heat resistant, the foam quickly coats the fuel itself and deprives it of oxygen. It literally smothers the fire, and then prevents the fuel from reigniting. It has the name "film forming foam" since that is exactly what it does.

Think for moment about jet fuel. AFFF snuffs out a jet fuel fire in mere seconds when using the proper technique but that technique has to be taught and practiced.

Put simply, extinguishing a jet fuel fire with AFFF requires training, lots of training. Who are the big consumers of jet fuel? Each and every airport in the world and every military installation with aircraft.

It is the adage "practice makes perfect" that perhaps lies at the genesis of PFAS proliferation. The military, including the Air Force, Navy, Marines, and Coast Guard, loves to train and loves even more to train with a purpose. Training to put out a jet fuel fire, therefore, is encoded into the military's DNA. With that in mind, it should come as no surprise that, at the time of this writing, over two hundred military installations (including nearly all Air Force bases active and deactivated) have elevated levels of PFOS in the ground and drinking water. The military trained using AFFF, the AFFF flowed freely into the ground and eventually into the water surrounding bases.

Military personnel using Aqueous Film Forming Foam (AFFF) in a training exercise. https://www.military.com/daily news/2016/09/12/firefighting-foam-linked-water-contamination-injuries-fire.html

The military is in a particularly vulnerable position when it comes to PFOS; it is both the polluter and the victim. The armed forces are wholly responsible for allowing the PFOS-laden AFFF to seep into the soil and then the groundwater and at the same time its very own workforce – the enlisted men and women – were the ones first exposed to the AFFF.

Wurtsmith Air Force Base, decommissioned in 1993, sits amid an idyllic setting on the shores of Lake Huron in east-central portion of the lower peninsula of Michigan. Its primary function throughout its seventy-year history was that of a combat crew and bomber training base. Where there's combat and bomber crews being trained there are also firefighters being trained to be able to handle the unthinkable should it ever arise. As has already been established, where there are firefighters in the Air Force there is training and where there is Air Force firefighting training there is AFFF and PFOS.

The accumulative results of all that training can be found in the data collected by State of Michigan:

> *"Fifteen foam samples have been taken off-base. A foam health advisory was issued by the local health department. Results range from 2,237 - 110,830 ppt PFOA+PFOS; and 2,487 - 164,253 ppt total tested PFAS. Cedar Lake (off base) foam results at 165 ppt PFOA and PFOS."*[86]

Present EPA allowable levels of PFOA and/or PFOS are 70 ppt. The lowest of those test results is more than twice the allowable limit while the other readings are incomprehensible. It should come as no surprise that multiple lawsuits have been and

---

[86] https://www.michigan.gov/pfasresponse/0,9038,7-365-86511_82704_83952---,00.html

are currently being litigated in Michigan based on the Wurtsmith pollution alone.

That's just one military base. As of this writing there are 283 military installations where PFAS readings exceed – and in the case of a base like Wurtsmith, far exceed – EPA allowable levels. Here is a table of those bases by state:

AK9	KY1	NV1
AL3	LA3	OH7
AR2	MA7	OK6
AZ12	MD14	OR12
CA28	ME6	PA5
CO4	MI8	RI4
CT1	MN2	SC5
DC1	MO7	SD4
DE2	MS9	TN6
FL15	MT4	TX14
GA5	NY11	UT3
HI0	NC4	VA6
IA3	ND5	VT5
ID2	NE5	WA4

IL 3	NH 3	WI 5
IN 3	NJ 7	WV 1
KS 1	NM 2	WY 3

The PFOS in the great majority of all AFFF manufactured came from 3M. At the same time, they were manufacturing the PFOS as an ingredient for AFFF in other manufacturer's products, they were manufacturing their own firefighting product,

Light Water® which also contained PFOS as a key ingredient. As we'll find out shortly, 3M did eventually phase out the version of PFOS found in firefighting foam in 2002, forty years after its inception in the mid-1960's. Forty years after all that training at all those bases, all that training by airport fire departments around the world had soaked the ground and contaminated the drinking water with PFOS.

There are those who contend AFFF is the true culprit here. Its extensive use in the military and at airports means that thousands of gallons of PFOS freely found its way into the ground and into the water; the argument carries a lot of validity. If it is true, it is yet one more front where 3M faces product liability litigation. This front, the AFFF front, could end up being far more crucial in that it affects the men and women in the United States military, those who fight for our nation who now have another battle to wage.

Response from the military has been swifter than that of EPA. The Department of Defense released this statement:

> *On January 28, 2016, the
> Department of Defense issued a policy
> requiring Military Service-specific risk
> management procedures to prevent
> uncontrolled land-based AFFF releases
> during maintenance, testing, and training
> activities. Additionally, the policy requires
> the DoD Components to remove and
> properly dispose of AFFF containing PFOS
> from the local stored supplies for non-
> shipboard use, where practical.*[87]

The Air Force, citing uses at active bases, reserve, Air National Guard and closed bases stated on March 16, 2016:

> *To prevent releases of firefighting
> foam in the future, the Air Force limits use
> of the foam to emergency responses only,
> and in those situations immediate action is
> taken to ensure containment. The Air Force
> is committed to eliminating firefighting
> foam containing either PFOS or PFOA from
> its inventory, and is finalizing a phased plan
> to replace existing firefighting foam
> inventories with recently approved
> PFOS/PFOA-free alternatives that still
> provide adequate fire protection for critical
> assets and infrastructure. These
> alternatives do contain PFCs but do not
> contain the two addressed by the EPA
> advisory.*[88]

---

[87]

https://www.denix.osd.mil/derp/home/documents/alternatives-to-aqueous-film-forming-foam-report-tocongress/

[88]

https://www.afcec.af.mil/Portals/17/documents/Environment/AF

The Navy then took their own stand:

> *In response to concerns over PFOS and PFOA, the Department of the Navy (DON) 1 amended MIL-PRF-24385F in 2017. The amendment identifies DoD's goal to develop and transition to a non-fluorinated agent and encourages AFFF manufacturers minimize the levels of PFOS and PFOA in their products in the interim. The amendment established a maximum concentration for PFOS and PFOA at the limit of quantitation of current test methods (800 parts per billion (ppb)). Due to the complex chemical matrices involved in analyzing per- and polyfluoroalkyl substances (PF AS) 2 in finished AFFF concentrates, DON is working with industry leaders to develop a standardized PF AS test method for lower quantitation levels.*[89]

And finally, just last year, the Army said:

> *The Army plans to replace its supply of long chain (C8) AFFF which contains PFOS and PFOA, with shorter chain (C6) AFFF, on the DoD qualified product list (QPL). C8 AFFF refers to the AFFF formulations containing long-chain*

D-160322-009.pdf

[89]

https://www.denix.osd.mil/derp/home/documents/alternatives-to-aqueous-film-forming-foam-report-tocongress/

> *fluorosurfactants, where long-chain is defined as seven or greater carbons, that DoD previously used and is removing from its inventory. Long-chain fluorosurfactants do not break down in the environment, spread rapidly, and bioaccumulate (i.e., become concentrated in humans). C6 AFFF refers to new AFFF formulations containing short-chain fluorosurfactants, where short-chain is defined as six or fewer carbons. The Army will dispose of the C8 AFFF and other AFFF-related waste3 by incineration. The AFFF waste disposal contracts are scheduled for award in August 2018.*[90]

When it comes to the damage already done to our military men and women, the Veterans Administration lags behind the others. The VA says this about PFAS:

> *"If you are concerned about health problems associated with exposure to PFAS during your military service, talk to your health care provider or local VA Environmental Health Coordinator. Veterans may file a claim for disability compensation for health problems they believe are related to exposure to chemicals during military service. VA decides these claims on a case-by-case basis."*[91]

The vast majority of PFAS claims submitted to the VA by veterans have been declined.[92] The can of worms remains unopened.

---

[90]

https://www.denix.osd.mil/derp/home/documents/alternatives-to-aqueous-film-forming-foam-report-tocongress/

[91] https://www.publichealth.va.gov/exposures/pfas.asp

CHAPTER SEVEN

1998: Emerging
into the Light of Truth

"Lies are convenient when the truth is unfathomable."
— Courtney M. Privett

From January 10, 1950 to May 15, 1998 is 17,657 days. Forty-eight years, four months and five days. Forty-eight years, four months and five days is longer than the lifetimes of John Lennon, Edgar Allan Poe, Glenn Miller, Jane Austen, Bobby Kennedy, John Kennedy, Elvis Presley and F. Scott Fitzgerald. Forty-eight years, four months and five days is also the amount of time between the first internal 3M toxicity test on mice in January of 1950 that showed proof of PFAS's bio-persistence and the date when 3M first disclosed to the EPA "the presence of organic fluorochemicals in the blood of the general population"[93] on May 15, 1998. Forty-eight years, four months and five days of tests both internal and external that found PFAS in nearly every living creature tested; polar bears at the

---

[92]   https://www.mlive.com/news/2019/11/michigan-veterans-face-uphill-battle-proving-toxic-exposure.html

[93] Exhibit 2602, State of Minnesota v. 3M Co., Court File No. 27-CV-10-28862

North Pole, penguins at the South Pole, sea eagles in the Baltic Sea and pelicans in Miami and Nevada. Forty-eight years, four months and five days of non-disclosure. Forty-eight years, four months and five days of delaying research into the proliferation, remediation and cure for PFAS's bio- and eco-persistence. Forty-eight years, four months and five days that could have and should have been put to use stopping global PFAS contamination. If guilty of nothing else, 3M's lack of timely disclosure delayed what might have been a preventable environmental justice catastrophe from being inflicted on planet Earth; we'll never know what might have happened had that been the case.

Curiously absent from the twenty-seven million pages of discovery documents in the 2010 Minnesota lawsuit is one that says "after forty-eight years, four months and five days we probably should tell the authorities responsible for the Toxic Substances Control Act that we have known about the toxicity of a substance we have manufactured since the 1950's" since that would have been an admission of guilt. Something did finally start a chain of events that would begin to shed light on the secrets held for so long; for forty-eight years, four months and five days to be exact.

The trail to revelation begins with a very intriguing series of internal 3M documents starting in early 1998. The first of those documents is what appears to be a presentation by a 3M Senior Environmental Engineer entitled "Environmental Fate and Effects of Fluorochemicals"[94]. Assigned the date of 1/1/1998 in the court documents, this presentation is mostly an overview of the chemical properties of fluorochemicals with very little historical insight. It is probably safe to assume this was a presentation to a technical type group or committee, the 3M Fluorochemical Steering Committee, perhaps. We know, again from court documents, that the FC Steering Committee, as it was called, met on January 19, 1998 and shortly after that meeting 3M became laser-focused on the issue of PFAS. It could be easily concluded this document was used in conjunction with that meeting as a presentation, a handout or background information prior to the meeting.

Although technical in nature, the presentation we're referring to does contain phrases like the ones found on page ten of the presentation, the page entitled "Environmental Issues of Fluorochemicals"[95]. Among the bullet points listed on that particular page are the phrases "Persistent", "Widely Dispersed in the Environment", "Potential to Bio-accumulate", "Biologically Active"[96]. In other words, all the things we now know about PFAS were being revealed to some group within 3M for what looks like the first time using nomenclature we now

---

94 Exhibit 2695, State of Minnesota v. 3M Co., Court File No. 27-CV-10-28862

95 Exhibit 2695, State of Minnesota v. 3M Co., Court File No. 27-CV-10-28862

96 Exhibit 2695, State of Minnesota v. 3M Co., Court File No. 27-CV-10-28862, pg. 10

commonly use when discussing PFAS more than twenty years later. It was also information 3M acquired during the past forty-eight years of dealing with PFAS.

When presenting "Light Water™ AFFF Usage Wastes" on page twenty-two of the presentation, the topics listed are "Continuing Issue with Customers", "Potential Environmental Harm", and "Potential Negative Publicity for the Corporation"[97]. The three points under "Potential Environmental Harm" are: "Fish kills", "Bird exposure", and "Disruption of Wastewater Treatment Plants"[98]. This page is subtle but damaging at the same time. It openly uses the word "continuing" when talking about issues with customers, the very definition of continuing being "ongoing". The problem with customers had been there a while and 3M now openly admitted it. What those specific customer problems were is not disclosed. They also admit to PFAS' injurious effect on fish the first example of that being, of course, forty-eight years prior.

This is a factual report, one where no recommendations are offered but admits in a technical manner there are continuing issues with customers, potential environmental damage and also damage to the 3M corporate image and reputation. These are the stakes in the game if the company would choose to finally acknowledge the issues related to PFAS contamination. What it does not reference is the known harm to humans. This is thirteen years after PFAS was first found in blood samples of the general public but that fails to get a mention in this report. It seems as though this presentation could have been the trigger point that started 3M thinking it had gotten way past the point of propriety; it seems as though

---

[97] Exhibit 2695, State of Minnesota v. 3M Co., Court File No. 27-CV-10-28862, pg. 22
[98] Exhibit 2695, State of Minnesota v. 3M Co., Court File No. 27-CV-10-28862, pg. 22

the time for honesty and disclosure had arrived; it seems as though some better angels within the company were issuing a clarion call to stop the deception. There were, however, corporate hoops to jump through, i's to be dotted and t's to be crossed before truth would finally see the light of day.

We need to take a quick mathematical side trip before we go any further. "Parts per" reporting has become standardized when discussing PFAS levels. All numbers are now based on parts per trillion or ppt. Seventy drops of PFAS in twenty Olympic-sized swimming pools would be equivalent to the EPA's acceptable level of 70 ppt. Some of the samples reported on in the various internal 3M documents use differing "parts per" numbers. As a clarifier, 70 parts per trillion is equal to 0.07 (one zero) parts per billion[99] and 0.00007 (four zeroes) parts per million[100].

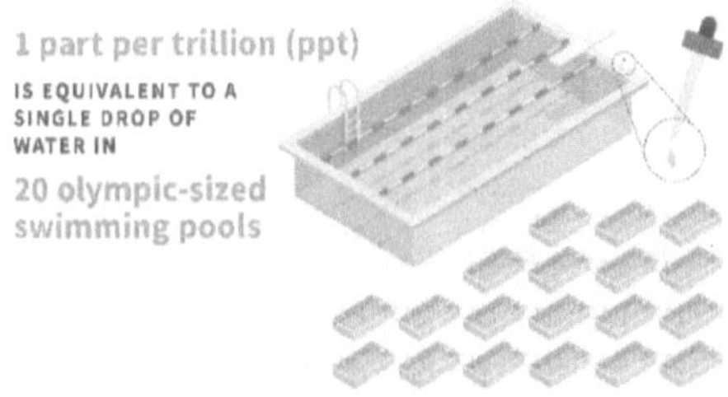

https://www.michigan.gov/pfasresponse/0,9038,7-365-86510_87918-474937--,00.html

Undated but also attributed to early 1998 is a more anecdotal report, relying less on data and more on descriptive verbiage. It is entitled "Fluorochemicals in

---

[99] https://www.easycalculation.com/unit-conversion/Parts_Per_Trillion-ppt-Parts_Per_Billion-ppb.html
[100] https://www.easycalculation.com/unit-conversion/Parts_Per_Trillion-ppt-Parts_Per_Million-ppm.html

Human Blood"[101]. This is a draft of a 3M internal document that reviews various tests and findings as far back as "1959"[102] It also references tests conducted in "February and April of 1998[103] so it fits that timeframe of those very intriguing series of internal 3M documents. Remember, 3M finally disclosed "the presence of organic fluorochemicals in the blood of the general population"[104] in a letter to the EPA on May 15, 1998 so this report, a report specifically about "Fluorochemicals in Human Blood"[105] almost certainly played a role in that decision to disclose. This report, this separate standalone report is what was missing from the previous report from early 1998. This report focuses solely on blood when there was not a mention of issues with human health in the other. This report fills in the gap and completes the story by delineating the other key issue with PFAS: its universal presence in human blood samples.

The language of this report is brutally honest. The second sentence in this fourteen page document says: "The presence of an organic form of fluorine in human blood was observed 30 years ago"[106] acknowledging in writing that 3M was aware of PFAS in human blood as early as far back as 1968. Early on in the report it also deceptively discloses PFAS's tendency to bio-accumulate when it says "(a)ge was

---

[101] Exhibit 1479, State of Minnesota v. 3M Co., Court File No. 27-CV-10-28862

[102] Exhibit 1479, State of Minnesota v. 3M Co., Court File No. 27-CV-10-28862, pg. 9

[103] Exhibit 1479, State of Minnesota v. 3M Co., Court File No. 27-CV-10-28862, pg. 12

[104] Exhibit 2602, State of Minnesota v. 3M Co., Court File No. 27-CV-10-28862

[105] Exhibit 1479, State of Minnesota v. 3M Co., Court File No. 27-CV-10-28862

[106] Exhibit 1479, State of Minnesota v. 3M Co., Court File No. 27-CV-10-28862, pg. 1

significantly associated with increased serum PFOS"[107]; the older you are, the more PFAS in your system. The human body just doesn't know how to metabolize PFAS substances.

Key to the document, is the attention given to nine global human serum tests between 1951 and 1994:

- *Korean War era military recruits, 1948-1951 – All were below 1 ppb*

- *Swedish samples, 1957 – The mean was 1.98 ppb (acceptable EPA level is 0.07 ppb)*

- *Michigan Breast Cancer Study, 1969 – 1971 – The mean was 33.4 ppb*

- *Swedish samples, 1971 – The mean was 1.13 ppb*

- *MRFIT[108] pooled calibration samples, 1976 – The mean was 30.9 ppb*

- *MRFIT pooled calibration samples, 1980 – The mean was 25.5 ppb*

- *China samples, 1984 – all below limits of detection*

- *MRFIT individual samples, 1985 – the mean was 30.7 ppb*

----

[107] Exhibit 1479, State of Minnesota v. 3M Co., Court File No. 27-CV-10-28862, pg 3

[108] The Multiple Risk Factor Intervention Trial (MRFIT). A national study of primary prevention of coronary heart disease. (1976)

- *China samples, 1994 – below limits of detection*[109]

      The comment in the report regarding the data listed above is the levels "do not appear to have changed significantly in the last 25 to 30 years"[110], a classic case of corporate spin. Ignoring the fact that the levels exist in the first place, the data in the report tells a different tale.

      The report says there were no discernible levels of PFAS in army recruit blood samples in the early 1950's; that then becomes the baseline. Sweden was already above the acceptable EPA parts per billion level by 1957 and in 1971 blood levels in Michigan, a hotbed of PFAS contamination, averaged – averaged – 33.4 parts per billion. Once more as a point of reference: "the U.S. Environmental Protection Agency has set a health advisory level of .07 ppb, or parts per billion, for the PFC chemicals PFOA and PFOS"[111]. What is staggering about that statistic is that we went from zero to sixty or, in this case, 0 to 33.4 ppb in less than twenty years. And, that was the mean number. The high-end number in that particular 1969 – 1971 Michigan sampling was 59.4 ppb. Acceptable level: 0.07 ppb; high-end in the 1971 Michigan test, 59.4 ppb. According to this report, with the exception of the Korean War samples and the ones from China, all sampled serum contained levels of PFAS that far exceeded EPA standards. The first PFAS production had barely begun while the Korean War raged and there is a bit of spin going on with the wordplay used to describe the results from China. The report uses the phrase "below

---

[109] Exhibit 1479, State of Minnesota v. 3M Co., Court File No. 27-CV-10-28862, pgs. 5-6

[110] Exhibit 1479, State of Minnesota v. 3M Co., Court File No. 27-CV-10-28862, pg. 6

[111] https://ohiovalleyresource.org/2016/10/21/chemicals-found-water-means/

limits of detection"[112]. That doesn't mean it wasn't there, it was just not detectable utilizing 1984 and 1994 technology in the words of the report.

Those Chinese reports, however, are a smokescreen. In the database of the discovery documents from the Minnesota lawsuit is document number 1195, a piece of 3M Internal Correspondence entitled "Fluoride Analysis of China Serum (AT1011)"[113] dated February 6, 1979. "Eight human serum samples were received from donors understood to live in rural China"[114] the memo begins and then goes on to say "(o)ne would conclude that there is a very low level of organic fluoride present."[115] The serum samples were measured in parts per million so, again, an acceptable level from the EPA now is 0.00007 parts per million. The organic levels for each of the eight samples was listed as:

- *Person 1 – 0.008 ppm*

- *Person 2 – 0.013 ppm*

- *Person 3 – 0.011 ppm*

- *Person 4 – 0.014 ppm*

- *Person 5 - 0.009 ppm*

---

[112] Exhibit 1479, State of Minnesota v. 3M Co., Court File No. 27-CV-10-28862, pg. 5

[113] Exhibit 1195, State of Minnesota v. 3M Co., Court File No. 27-CV-10-28862

[114] Exhibit 1195, State of Minnesota v. 3M Co., Court File No. 27-CV-10-28862

[115] Exhibit 1195, State of Minnesota v. 3M Co., Court File No. 27-CV-10-28862

- *Person 6 – 0.009 ppm*

- *Person 7 – 0.004 ppm*

- *Person 8 – 0.017 ppm*[116]

By 2020 EPA standards, all of these samples far exceed acceptable levels. Two other items about this particular sampling: first, this is rural China in 1979, less than thirty years after PFAS was first produced yet these rural Chinese were contaminated with PFAS beyond acceptable levels; secondly, and more importantly, 3M didn't disclose this report even in its own internal presentations. Thirty years after PFAS was first mass produced and almost twenty years before 3M disclosed the dangers it created, this memo on blood samples from rural Chinese was overlooked. It also begs us to take a closer look at the two Chinese studies cited (1984 and 1994) and it seems suspicious that no specific data is associated with either of those two samplings. Where were the 1979 findings and, if there were specific values for 1979, why wouldn't there be the same five and fifteen years later? The 1979 memo contains this confusing phrase: "past samples of Red Cross plasma have analyzed to contain a trace level of FC-95 but we are not sure if the level seen is truly FC-95"[117]. This statement seems to be saying it's there but we're not sure if it's there. An argument to itself, one would question where else a fluorocarbon like FC-95 would come from. Then, in the very next sentence it says "By

---

[116] Exhibit 1195, State of Minnesota v. 3M Co., Court File No. 27-CV-10-28862

[117] Exhibit 1195, State of Minnesota v. 3M Co., Court File No. 27-CV-10-28862

subjecting the Chinese samples to the same analytical scheme, one should be able to establish if we can conclude the presence of 0.01 ± 0.01 ppm FC-95 in Red Cross plasma" which continues to confuse the reader and leaves one to wonder did they really test these samples at all. The conclusion is they tested the samples and all indications were that organic fluorocarbons were present, but they just weren't really sure if they could specifically call it FC-95. How else could there be a fluorocarbon (a manmade substance) presence in the blood of rural Chinese unless it came from PFAS contamination?

3M's claim that samples "do not appear to have changed significantly in the last 25 to 30 years"[118] rings false unless 3M has a completely different definition of "significant".

Despite the fact both sets far-exceeded acceptable EPA levels, the mean number from the 1980 MRFIT samples were actually greater than the ones from 1976. These were also mean numbers and those mean numbers could have been drastically affected by where the samples came from. 3M hides behind the mean. The report itself claims "(t)he location means ranged from 14 ppb in Santa Barbara, California to 52 ppb in Greenville, South Carolina"[119], so

---

[118] Exhibit 1479, State of Minnesota v. 3M Co., Court File No. 27-CV-10-28862, pg. 6
[119] Exhibit 1479, State of Minnesota v. 3M Co., Court File No. 27-CV-10-28862, pg. 5

where the MRFIT samples came from (undisclosed in the report) would have a major impact on the mean numbers.

The Toxic Substances Control Act is a federal law passed by Congress in 1976. This is what section 8(e) of that law says:

> **_NOTICE TO ADMINISTRATOR OF SUBSTANTIAL RISKS._** _Any person who manufactures, [imports,] processes, or distributes in commerce a chemical substance or mixture and who obtains information which reasonably supports the conclusion that such substance or mixture presents a substantial risk of injury to health or the environment shall immediately inform the [EPA] Administrator of such information unless such person has actual knowledge that the Administrator has been adequately informed of such information. -- Section 8(e), Toxic Substances Control Act (1976)[120]_

You should notice two things: 1) TSCA was passed by Congress in 1976; and 2) TSCA applies to new chemicals only. You should see what's coming next like a 747 through your front door: PFOA and PFOS were both invented prior to 1976 and, therefore, are not covered by TSCA. Or at least that was 3M's interpretation. 3M felt it didn't need to tell the EPA is was manufacturing a chemical that posed a risk to man and the planet. Their logic was since PFOA and PFOS pre-dated TSCA there was no legal requirement to report but there certainly was, as anyone with a moral compass would agree, an obligation to disclose. That, however, remained a problem with 3M, for they still, in early 1998, had not reported the toxicity it had discovered back in 1950.

---

[120] https://www.epa.gov/sites/production/files/2015-09/documents/1991guidance.pdf

3M, like many other corporate entities, loves its layers of authority. Supervisors report to managers who report to directors who report to vice presidents, and on and on. An office rather than a cubicle is the first step then it becomes office size, windows and office location that matter. When asked why the CEO had never stepped in to do the right thing, the sad answer is the various CEOs who led the company between 1950 and 1998, four in all, probably had no clue as to the extent of what was going on. The company insulates its upper management from the mundane everyday details by assigning power to the individual operating units. The power is spread throughout the company in a maze of divisions and departments. There is evidence in the public documentation that presentations were made to the Board of Directors (of which the current CEO is always a member) down through the years but those presentations were always examples of the mastery of spin 3M had attained: never quite a lie but never the whole truth about PFAS.

3M also likes its committees and so it should be no surprise that the company had and probably still has a TSCA (Toxic Substances Control Act) Committee. That committee met on Friday February 13, 1998, a Friday the thirteenth in a foreboding piece of irony. No record of that meeting exists, at least not in the discovery documents, yet we know it happened since it is referenced in the notes from a meeting held thirteen days later on Thursday February 26, 1998. Under the heading "EPA COMMUNICATION PLAN"[121], the first point in the presentation from that second meeting says "Corporate TSCA committee met on 2/13/98 … Consensus to file 8(e)[122] to EPA".[123] In plain English, after

---

[121] Exhibit 1488, State of Minnesota v. 3M Co., Court File No. 27-CV-10-28862, pg. 3

[122] Section 8(e), the substantial risk information reporting provision of the Toxic Substances Control Act (TSCA). - https://www.epa.gov/assessing-and-managing-chemicals-under-

nearly fifty years of avoidance, 3M had finally decided to tell the federal government that PFAS was present in the bloodstream of the general population. Finally. This communication plan called for the letter to be sent "on or before 3/6/98"[124]. Not so fast.

In a letter dated March 20, 1998 and marked "3M Confidential" "Do Not Copy" an abrupt halt was put on the consensus to disclose by a high-ranking 3M official. "It is 3M's policy to comply with the Toxic Substances Control Act, including the reporting requirements of TSCA section 8 (e)"[125] the letter begins and then the author admits to conducting a *de novo* review, a nonreferential standard of review[126]. Based on what the author discovered in that review, the last paragraph of the letter says in part, "I have concluded that 3M is not presently in possession of information that would be new to EPA and that reasonably supports a conclusion that suggests a substantial risk of injury to human health or the environment"[127]. Dead fish, mice, rats, birds, goats and monkeys to the contrary. Blood samples from around the world to the contrary. Polar bears and penguins to the contrary. This executive was attempting to, singlehandedly, disrupt and countermand the work of all those honest 3M employees who saw the past and feared the future. One man with profits as a

---

tsca/tsca-section-8e-reporting-guide

[123] Exhibit 1488, State of Minnesota v. 3M Co., Court File No. 27-CV-10-28862, pg. 3

[124] Exhibit 1488, State of Minnesota v. 3M Co., Court File No. 27-CV-10-28862, pg. 3

[125] Exhibit 1496, State of Minnesota v. 3M Co., Court File No. 27-CV-10-28862

[126] https://www.investopedia.com › terms › de-novo-judicial-review

[127] Exhibit 1496, State of Minnesota v. 3M Co., Court File No. 27

motivator trying to reverse the decision of those previously mentioned better angels.

This very intriguing series of internal 3M documents continues, the next one an almost full-frontal attack on the executive decision to not file a TSCA 8(e). Dated May 1, 1998, exactly six weeks after the reversal of the decision to file allowing plenty of time to gather all of the pertinent data contained in it, the letter clearly drives home the point that the committee was right, and the executive was wrong about filing with the EPA. Redacted portions aside, the letter – more of a report, really – argues "Perfluorochemicals are persistent" (page 2), "some fluorochemicals have significant biological activity" (page 3), "fluorochemicals can bioaccumulate" (page 3), and "fluorochemicals are being released to the environment" (page 3)[128]. This report completely refutes the executive's decision and says, yes, fluorochemicals and POSF and PFOS in particular present a "substantial risk". It is an in-your-face rebuttal to an executive who, being first accountable to stockholders, seemed more concerned about an income stream than doing the right thing.

The good guys finally won, albeit forty-eight years, four months and five days too late. 3M finally told the EPA on May 15, 1998 that the "presence of organic fluorochemicals in the blood of the general population and subpopulations, such as workers, has been known dating back to the 1970's"[129]. In a final bit of irony, the letter was signed by the very executive who tried to put a halt to the filing less than two months earlier.

----

[128] Exhibit 1502, State of Minnesota v. 3M Co., Court File No. 27
[129] Exhibit 2602, State of Minnesota v. 3M Co., Court File No. 27-CV-10-28862

The letter is also the first time 3M employs the strategy of "a little bit of PFAS won't hurt anybody". "3M does not believe that any reasonable basis exists to conclude that PFOS 'presents a substantial risk of injury to health or the environment'"[130] the quote within the quote being the wording from section 8(e) itself, an in-your-face at EPA. It permeates the planet yet there is no substantial risk. Contained in blood samples of the general population yet there is no substantial risk. In surface water, groundwater, fish, birds and plant life yet there is no substantial risk. From the first moment of self-disclosure the 3M tactic has been there is no reasonable basis to conclude that this bio-accumulative, bio-persistent, eco-persistent organic fluorochemical they created which saturates the planet poses any substantial risk. That would be a defense that in the years to come, quite literally, would not stand up in court.

The delay in reporting would be costly to 3M but, perhaps not as costly as it should have been. Eight years almost to the day after it finally reported PFAS under section 8(e) of the Toxic Substances Control Act, EPA released the following statement highlighting what could be called a mere slap on the wrist for such an egregious error:

> *(Washington, D.C. - April 25, 2006) EPA and the 3M Company reached a $1.5 million settlement to resolve reporting violations under the Toxics Substances*
> *Control Act (TSCA) that the company voluntarily disclosed to EPA. EPA filed the settlement with the agency's Environmental Appeals Board today, for their review.[131]*

---

[130] Exhibit 2602, State of Minnesota v. 3M Co., Court File No. 27
[131] https://www.epa.gov/enforcement/3m-company-settlement

It would be another four years before 3M stopped manufacturing those forms of PFAS, PFOS and PFOA. Four more years of profits, four more years of proliferation.

CHAPTER EIGHT

The Year 2000 – Tap Dancing Around the Truth

"This isn't a health issue now, and it won't be a health issue"[132] – Larry Zobel, 3M Medical Director

"ST. PAUL, Minn -- May 16, 2000 -- 3M today announced it is phasing out of the perfluorooctanyl chemistry used to produce certain repellents and surfactant products."[133]

---

132 David Barboza, "E.P.A. Says it Pressed 3M for Action on Scotchgard Chemical", *New York Times*, May 19, 2000
133 Exhibit 1690, State of Minnesota v. 3M Co., Court File No. 27-CV-10-28862

So began the Press Release announcing, slightly more than two years after notifying the EPA of the risks associated with PFAS products, 3M was "phasing out" of their manufacture. By April of 2000 the preponderance of evidence was not in 3M's favor. PFAS in children's sera, Baltic Sea eagles along with streams and rivers around the world. This would be the first time the general public heard of PFAS but, as has become a recurring theme, it would not be the truth, the whole truth and nothing but the truth.

"Sophisticated testing capabilities -- some developed in only the last few years -- show that this persistent compound, like other materials in the environment, can be detected broadly at extremely low levels in the environment and in people. All existing scientific knowledge indicates that the presence of these materials at these very low levels does not pose a human health or environmental risk."[134] This quote from the release is spin at its zenith. The last sentence, especially, tells the truth in a highly filtered manner something which the period from 1998 through 2000 and the years beyond ooze with. That sentence is absolutely true but if you focus on three words – "very low levels" – you begin to see the spin. Yes, PFAS does not pose a risk to human health or the environment at "very low levels", very few things do. However, that statement does not take into account the bio-accumulative effects of PFAS.

The "very low levels" you ingest today do not "pose a human health or environmental risk". But added to the "very low levels" you ingest tomorrow and to the ones from next week and next month and next year, the bio-accumulative tendencies of PFAS mean you no longer have "very low levels" you now have substantial if not significant

---

[134] Exhibit 1690, State of Minnesota v. 3M Co., Court File No. 27-CV-10-28862

levels. Many doses of "very low levels" add up to a lot over time since they never go away; it is, after all, the "forever chemical".

Secondly, in some existing cases of PFAS exposure we're not talking about "very low levels" we're talking about levels way beyond acceptable, way beyond legal. Remember the church down the road from the 3M Wood bury dumpsite from Chapter Five? Its PFAS levels were 600 ppt or 8.5 times the acceptable EPA level and more than 16.5 times greater than the allowable level in the State of Minnesota. That is not a "very low level". That statement in the press release is purposely misleading. Now that there had been disclosure to both the EPA and the general public, damage control through spin was becoming the order of the day at 3M. The company would now deploy defense tactics which many experts categorize as identical to those used by other industries under siege such as tobacco and pharmaceuticals.

The period between 1998, when 3M first reported PFAS' presence in humans to the EPA for the first time, and 2000 when they announced to the world they were "phasing out" of PFAS production was a time laced with more twists, turns, subplots, betrayals, revelations and back-peddling than in the entire prior history of the company. The tide of conventional wisdom was not in their favor. Their corporate ethos had turned into pathos. The hallowed, respected halls were being dimmed by what most certainly was a classic case of corporate irresponsibility. The few who knew down through the years had wrought a present and future of grave concern.

PFAS spin became known around 3M as "Command the Science" and the company drew from the halls of academia to fill that need, not for research and science but for control of the facts. Enter John P. Giesy. In 1998, John Giesy was a zoology professor at Michigan State University in East Lansing, Michigan. Among his achievements, Giesy is credited with being the first to discover ecological evidence of PFAS proliferation. That was how his relationship with 3M began. As of this writing, Giesy is a Canada Research Chair at the University of Saskatchewan in Saskatoon, Saskatchewan, Canada. In a March 2018 news article, the Canadian Broadcasting Company – CBC – traced the questionable relationship between Giesy and 3M. The data in the article, and also subsequent news items by Minnesota Public Radio, the *Sydney Morning Herald* in Sydney, Australia and M Live in Michigan were all based on allegations filed by the State of Minnesota in November of

2017. 

Shortly after disclosing PFAS to the EPA, "Professor Giesy's consulting company appears to have received payments from 3M between at least 1998 and 2009. One document indicated his going rate was about $US275 an hour."[135] "'Despite spending most of his career as a professor at public universities, Professor Giesy has a net worth of approximately $20 million,' states a memorandum to amend the complaint filed by Minnesota state lawyers

---

[135] Carrie Fellner, "Toxic Secrets: Professor 'bragged about burying bad science' on 3M chemicals", *The Sydney Morning Herald*, June 16, 2018

last November (2017)."[136] "This massive wealth results at least in part from his long-term involvement with 3M for the purpose of suppressing independent scientific research."[137]

That same memorandum from the State of Minnesota claims that Giesy:

- "Received more than $2 million in grants and payments from 3M and internally described himself as a member of the 3M team while portraying himself as an 'independent' editor of academic journals. Lists of payments are included in the documentation.
- Concealed the nature of certain payments from 3M. In a 2008 email, Giesy stated: 'I always listed these reviews as literature searches so that there was no paper trail to 3M.'
- Shared confidential academic manuscripts about PFCs with 3M employees.
- Worked with 3M to prevent 'bad' academic papers from being published. In another 2008 email, Giesy wrote that '...in litigation situations, they can be a large obstacle to refute.'
- Worked on 3M's behalf to 'buy favours' from other academics."[138]

What is even more distressing is, in a detailed email response to CBC News, Giesy:

---

[136] Jason Warick, "U of S professor denies suppressing toxic pollution research for 3M", CBC, March 11, 2018
[137] Jason Warick, "U of S professor denies suppressing toxic pollution research for 3M", CBC, March 11, 2018
[138] Jason Warick, "U of S professor denies suppressing toxic pollution research for 3M", CBC, March 11, 2018

- "Did not dispute the amount he received in grants and payments. He said he worked with 3M on replacement products for PFCs and also worked with a private company affiliated with testing the products.
- Said he isn't sure why he made the 'no paper trail' reference. He said it may have been related to the Chinese government collecting data on companies. 'I was not in the business of trying to suppress any information. Quite the contrary.'
- Said he did share a submitted academic manuscript with a 3M scientist while serving as a journal editor. Giesy said the measurements in the paper appeared incorrect. 'I made the decision to ask for an opinion from the other person in the world who knew how to do the analyses of (chemical) PFOS correctly. This was completely in my powers as the editor.'

- Said he did not suppress any academic papers to protect 3M from legal action. 'I published all of the information in the peer-reviewed open literature' including evidence their chemical compound was a 'cancer promoter.'

- Said he used the term 'buy favours' as an 'innocent turn of phrase.' Giesy said it referred to his efforts to get 3M to help China test for PFCs in its environment. 'We were not bribing anyone. No money was exchanged.'[139]

---

[139] Jason Warick, "U of S professor denies suppressing toxic pollution research for 3M", CBC, March 11, 2018

This is the man 3M paid millions of dollars to protect millions if not billions of dollars in business. A man who freely admits being paid millions and is so consumed by money that he calls the term "buy favors" an "innocent turn of phrase" and thinks that bribery is limited only to money being exchanged. To paraphrase Florence King, "people are treacherous for the simple reason that treachery is both a means of survival and a way to curry favor."

"University of Manitoba academic ethicist Arthur Schafer says the allegations against University of Saskatchewan professor John Giesy are the most serious he's ever seen. Schafer said if the allegations are true, it's the most 'extensive information laundering campaign' he's ever seen."[140] "Information laundering" or "Command the Science". You decide.

From the time EPA received the letter of May 15, 1998 they had grave concerns about the truth, the whole truth and nothing but the truth about 3M and its knowledge of the traits and tendencies of PFAS. They were completely reliant on 3M to tell the truth. Since PFAS was not – and still is not – classified as a toxic substance, there was no obligation on the part of 3M to disclose information about the chemical. This was self-disclosure. Failure to report the toxic substance, however, would be investigated and 3M, as noted previously, would eventually be fined $1.5 million for failure to do so but, from the point of disclosure, the EPA had a gnawing feeling facts were missing.

That feeling would be confirmed on Mach 28, 1999 when 3M Environmental Specialist Richard Purdy resigned his position with the company and sent a copy of his resignation letter to the EPA. The repercussions to that

---

[140] Jason Warick, "U of S professor denies suppressing toxic pollution research for 3M", CBC, March 11, 2018

letter resonated from Washington to St. Paul and back again. The term "whistle-blower" becomes a positive one in Purdy's case: he helped confirm what the Environmental Protection Agency suspected as early as 1998. In December of 1998 there was a proposed meeting between the agency and 3M to clear the air and "scope the issues surrounding the discovery of PFOS"[141]. About a month before 3M announced the phase out, there are the handwritten notes of a phone conversation between two high-level EPA officials that sets the tone for the 3M/EPA relationship and gives this work its title. The entry simply says "what was known and when"[142] a phrase dripping with suspicion. It seems EPA was beginning to feel they had been duped by 3M and wanted to confirm what the company knew, when they knew it and what lead to their failure to disclose.

Purdy's resignation letter is a full two-page accusatory epistle of all that 3M did and didn't do regarding PFAS. These are key points and highlights:

- "Perfluorooctansefulfonate (PFOS) is the most insidious pollutant since PCB. It is probably more damaging than PCB because it does not degrade, whereas PCB does."[143]

- "For more than twenty years 3M's ecotoxicologists have urged the company to allow testing to perform an ecological risk assessment on PFOS and similar chemicals. Since I have been assigned to the problem a year ago, the company has continued its hesitancy."[144]

---

[141] Exhibit 2719, State of Minnesota v. 3M Co., Court File No. 27

[142] Exhibit 1664, State of Minnesota v. 3M Co., Court File No. 27

[143] Exhibit 1001, State of Minnesota v. 3M Co., Court File No. 27-CV-10-28862

[144] Exhibit 1001, State of Minnesota v. 3M Co., Court File No. 27

- "3M submitted a TSCA 8e last May. There is tremendous concern within EPA, the country, and the world about persistent bio-accumulative chemicals such as PFOS. Just before that submission we found PFOS in the blood of eaglets--eaglets still young enough that their only food consisted of fish caught in remote lakes by their parents. This finding indicates a widespread environmental contamination and food chain transfer and probable bioaccumulation and bio-magnification. This is a very significant finding that the 8e reporting rule was created to collect. 3M chose to report simply that PFOS had been found in the blood of animals, which is true but omits the most significant information."[145]

- "3M waited too long to tell customers about the widespread dispersal of PFOS in people and the environment. We knew before May of 1998, yet 3M did not start telling customers until January of 1999. I felt guilty about this and told customers I personally knew earlier. Still, it was not as early as it should have been. I kept waiting for 3M to do its duty, as I was continually assured that it would."[146]

- "3M continues to make and sell these chemicals, though the company knows of an ecological risk assessment I did that indicates there is a better than 100% probability that perfluorooctansulfonate is biomagnifying in the food chain and harming sea mammals. This chemical is more stable than many rocks. And the chemicals the company is considering for replacement are just as stable and biologically available. The risk assessment I performed was simple, and not worst case. If worst

---

[145] Exhibit 1001, State of Minnesota v. 3M Co., Court File No. 27
[146] Exhibit 1001, State of Minnesota v. 3M Co., Court File No. 27-CV-10-28862

case is used, the probability of harm exceeds 100,000%."[147]

- "3M told those of us working on the fluorochemical project not to write down our thoughts or have email discussions on issues because of how our speculations could be viewed in a legal discovery process. This has stymied intellectual development on the issue, and stifled discussion on the serious ethical implications of decisions."[148]

- "I can no longer participate in the process that 3M has established for the management of PFOS and precursors. For me it is unethical to be concerned with markets, legal defensibility and image over environmental safety."[169]

Purdy's words speak for themselves and need no further embellishment. It is a view from the frontlines of PFOS research at 3M. It is the quintessential view from the inside.

Now that Purdy had blown the whistle "when EPA began making noise about possibly beginning the regulatory process for the 'grandfathered' PFOS chemical, 3M tried to preempt all of that by simply agreeing to yank PFOS off the market"[149]. And that's what they did; two years later. Tell

---

[147] Exhibit 1001, State of Minnesota v. 3M Co., Court File No. 27-CV-10-28862
[148] Exhibit 1001,
State of
Minnesota v. 3M
Co., Court File
No. 27 [169] Exhibit
1001, State of
Minnesota v. 3M
Co., Court File
No. 27

the EPA of its risks in 1998. Announce the "phase out" in 2000. Actually have the product in its original construction out of the marketplace in 2002. All of which allowed ample time for customers to stock up, one claiming over that two-year period having bought enough product to last ten years. Ten years of full-strength PFAS pollution.

In 2002, 3M introduced an improved version of PFAS. In Chapter Three, PFAS was characterized as having an eight-chain construction. The new and improved rendition of PFAS? A four-chain construction so instead of being C-8, it is now C-4, nothing more than an ersatz version of its predecessor. They are identical in every respect except for the length of the chain of chemicals that form its basic construction. If C-8 had been the "forever chemical", C-4 could now be called the "half of forever chemical". In 3M's eyes it was new, improved and less harmful to the environment. The damage it would cause would only last half of forever and that is a 100,000% certainty.

---

[149] Robert Bilott, *Exposure: Poisoned Water, Corporate Greed and One Lawyer's Twenty-Year Battle Against DuPont*
(New York, Atria Books, 2019) pg. 96

CHAPTER NINE

When Doing the Right Thing Loses Out to Self-Aggrandizement

"Corporate social responsibility is measured in terms of businesses improving conditions for their employees, shareholders, communities, and environment." - Klaus Schwab

If ever there was a shrine to college hockey, this would be it. On *Sports Illustrated's* 2007 list of the Top Ten venues in all of college sports it was the only hockey arena mentioned. It was listed as one of the five best college hockey arenas by LRT Sports as recently as 2018. As you pass through the gate and enter the building the sign over the door proclaims: "Through these gates walk the greatest fans in college hockey". These are people who aren't easily fooled when it comes to hockey or hockey arenas. The people passing through those gates call the place Mariucci Arena.

On a hot and breezy July morning in 2017 as the legal battle between the State of Minnesota and 3M plodded through its eighth year, a news story broke across the Twin Cities: "Maplewood-based 3M Co. paid $11.2 million for a 14-year sponsorship of Gophers Athletics, a deal that includes naming rights to the Gophers' hockey arena."[150] The press release from 3M later that day quoted their Chief Marketing Officer: "Our sponsorship of Mariucci Arena and Golden Gopher Hockey is a tribute to our Minnesota roots,"[151] at the very same time they were preparing to defend themselves for egregiously poisoning the non-metaphorical roots of each and every plant in a tiny portion of the same state of Minnesota. That very day the company that was balking at paying even a penny to clean up the toxic nightmare it had created in nine unsuspecting communities wrote a check for $11.2 million dollars just so it could have its name on a hockey arena. Yes, hockey is woven into the fabric of the State of Minnesota – the "State of Hockey" as its NHL franchise claims – but names on sports arenas should never take priority to the health and well-being of Minnesota's people or the global population.

150

https://www.bizjournals.com/twincities/news/2017/07/10/3m-buys-naming-rights-to-mariucci-arena.html
151

https://www.bizjournals.com/twincities/news/2017/07/10/3m-buys-naming-rights-to-mariucci-arena.html

http://www.citypages.com/news/mariucci-arena-has-a-dumb-new-name/433605443

    Somewhere on the road to doing the right thing, 3M took a wrong turn and wound up on a bramble-laden path where the blooms of honor and responsibility were being choked out by the weeds of ego and corporate narcissism. The $11.2 million was only the latest in a long line of sponsorships meant to draw attention to the company and its portfolio of products. Money that could have been earmarked for research grants to find a remediation solution for PFAS poisoning or perhaps as simple as being voluntarily invested in a clean-up effort in communities where thousands of their employees lived and raised their families was instead spent on self-promotion. Their corporate priorities meant spending money to have their logo on a hockey arena instead of purifying drinking water they knowingly polluted. As a side note, most local hockey fans still call it Mariucci Arena not the clumsy "3M

Arena at Mariucci" the $11.2 million was supposed to create.

They are called "200 mile-per-hour billboards" but the dollars spent on NASCAR team sponsorships are the lifeblood of the sport. NASCAR is a family-owned business not a publicly held company so trying to pin down the actual costs of sponsorship is like trying to grab a handful of Jell-O®. A commonly quoted sponsorship figure is between $5 million and $35 million per car per year. Then there are also the sponsors not tied to a specific team or car, those sponsors whose logos appear on each and every race participant. Those sponsorships, "in-kind" sponsorships, provide free product to all participating racing teams in exchange for logo placement on every Chevy, Ford and Toyota in the field.

https://www.sportspromedia.com/news/nascar_and_3m_extend_for_five_years

https://athlonsports.com/nascar/nascars-push-new-fans-running-out-gas

In 1995, long after 3M became aware of the dangers of PFAS but still before it openly admitted it to the world it had already polluted, the company began one of those NASCAR in-kind sponsorships and provided every team, every car, every NASCAR shop with "over 200 products"[152] each. Twenty-five years of supplying over 200

---

[152] https://www.bodyshopbusiness.com/3m-automotive-aftermarket-division-keeps-nascar-teams-fast-lane/

products to nearly four dozen race teams at no charge every year is a staggering sum the company is not willing to share. But in 2005, 3M raised the stakes and their investment to grandiose levels by becoming a primary sponsor of the number 16 Rousch Racing team car followed later by the sponsorship of the number 24 of Hendrick Motorsports shelling out that previously guesstimated $5 to $35 million dollars per year for a total of eleven racing seasons. Although actual figures are not available, it is safe to assume the minimal investment for both levels of sponsorship for the time periods involved greatly exceeds $100 million. And as the cars sped around the tracks, PFAS sped even faster throughout the global ecosystem. The score: NASCAR investment $100 million, 3M's investment in the clean-up and remediation of PFAS contamination, $0.

There was more. Back before the initial in-kind NASCAR sponsorship in 1995, 3M announced to employees, customers and the global community that it was to become a worldwide sponsor of the 1988 Olympics. Price tag: $15 million[153]. The global exposure for the company, a global company, was tremendous yet it came at a time when 3M knew it had caused global damage but had still not publicly admitted it. Fifteen million dollars for research or fifteen million dollars for self-promotion? Ego won.

---

[153] https://www.latimes.com/archives/la-xpm-1985-11-26-fi-2140-story.html

Four years later they invested at least another $15 million in the 1992 Olympics and four years after that another $15 million or more for the 1996 Olympics. A minimum of $45 million dollars for sponsorship of three Olympic Games four years before the public shame of their role in PFAS proliferation around the world appeared on the evening news. Spending $45 million for a global sponsorship instead of spending a like amount on global stewardship shows 3M's true corporate priorities. Their ego is valued higher than your health.

https://3mopen.com/

From the rattle and roar of NASCAR to the pastoral serenity and civility of the PGA, corporate sponsors pay millions to have their names and logos prominently featured. 3M is no exception investing generously in both NASCAR and the PGA. In addition to NASCAR, 3M first entered into a relationship with the PGA in 2001 as the name sponsor of the 3M

Championship on the PGA Senior Tour or as it was later renamed, the PGA Tour Champions.

Starting in 2019, the 3M Open began a stop on the regular PGA Tour. In 2007, a Senior Tour Sponsorship cost $1.8 million that escalated five percent each year through 2009. By the time 2009 rolled around the annual bill would have been nearly $2 million. If you use that $2 million as an average, the eighteen years of sponsorship of the Senior Tour event was $36 million and when you throw in another $10 million (title sponsorship costs in 2019 ranged from $8 million to $13 million so the average would be around $10 million) the total 3M spent on golf sponsorships was close to $50 million.

Added all together – the $11.2 million for naming rights at Mariucci Arena, an extremely conservative $100 million spent on NASCAR, $45 - $50 million for the Olympics and, finally, close to $50 million for PGA events – 3M spent in the neighborhood of a quarter billion dollars on primping and preening their corporate image while not spending a penny on PFAS research.

Warren Buffet once said "It takes 20 years to build a reputation and five minutes to ruin it. If you think about that, you'll do things differently."[154] Unless you are 3M. 3M spent nearly fifty years trying to build a reputation of honesty and environmental stewardship while all the time, hidden from the public view, they were doing the very things that could lead to destroying that reputation in Warren Buffet's famous five minutes. They didn't, as Buffet suggests, think about it. The only thought they had was to not reveal the truth to shareholders, government officials, employees and the general population.

Hop on Interstate 94 in front of 3M World Headquarters in Maplewood, Minnesota and head west. A short fifteen-minute drive – traffic permitting – puts you on the campus of a Tier 1 research university. In the top twenty institutions who receive Federal research dollars, this university that's down the road from 3M, places higher on that list than venerable institutions like MIT and Yale. While 3M was spending a quarter billion dollars on self-serving self-promotion, they didn't invest a single dollar on PFAS research with the University of Minnesota, just down the road. No money was invested in finding remediation for

---

[154] James Berman, "The Three Essential Warren Buffett Quotes to Live By", *Forbes*, April 20, 2014

an existing global problem. No money was even invested in finding an environmentally friendly, human friendly replacement for PFAS. Instead, 3M's priorities were race cars, Olympic venues and golf tournaments. Most galling is the fact that 3M spent $11.2 million to have its own name mounted on the University of Minnesota's men's hockey arena without spending any money on PFAS research on that same campus, that Tier 1 research campus, that highly acclaimed, readily accessible world class research facility that is the University of Minnesota, just down the road.

3M spent a quarter of a billion dollars on self-promotion during a period when they knew they were helping to proliferate a toxic substance around the world. It was all about 3M and creating subterfuge, a positive image that didn't match the behind-the-scenes corruption. 3M and its image came first, being honest with the world came second.

CHAPTER TEN

See You in Court? –
Consent Agreements, Lawsuits
and Settlements

*"They who pollute, sinned against nature." - Toba
Beta*

Assumed as one of the reasons for being less than forthcoming in its disclosure of the dangers of PFAS was 3M's fear of litigation. As has been said *ad nauseum*, we do indeed live in a litigious society. Delay, however, does not dispel and shortly after 3M finally began disclosing the inherent risks of PFAS, lawsuits began popping up far and wide. At last estimate, there are nearly 200 cases in process against the company for PFAS related issues six of which are being brought by states.

The initial damage was done in 3M's own backyard and the company has also felt the litigious damage close to home as well. Nearly one billion dollars have been awarded in Minnesota in three high profile cases. In brief summary, here are those three cases, what they were for, how much money is in play and what the actions mean in the long run. It should be pointed out that all of these cases were settled out of court and by agreement; no trial time has yet to be logged in Minnesota. Perhaps out of fear of what actions a jury might take, 3M has settled each of the following cases before trial.

<u>**2010 Settlement Agreement and Consent Order –
Minnesota Environmental Response and Liability Act
(MERLA)**</u>

The Minnesota Pollution Control Agency (MPCA) brought suit against 3M "for the purpose of providing for remedial investigations and response actions to address certain discharges to waters of the State and releases or threatened releases to the environment in order to minimize or abate pollution of waters of the State and to protect public health and welfare and the environment" based on "releases and threatened releases of PFCs at the 3M Cottage Grove Site, the 3M Oakdale Disposal Site and the 3M Woodbury Disposal Site." In other words, the MPCA wanted to have 3M clean up the messes it had created at three of the four SoWashCo dumpsites.

The SACO lists all of the offenses, site by site. The financial renumeration is to be paid to the cities involved for emergency and/or short-term solutions to PFAS-based drinking water issues. In total, some $40 million dollars is allocated to local contamination issues.

<u>**2018 Settlement**</u>

"In December 2010, the State of Minnesota filed a lawsuit against 3M Company seeking payment for natural resource damages caused by 3M's disposal of perflourochemicals, or PFCs, in the East Metropolitan Area of the Twin Cities in Minnesota. The lawsuit was brought by the Attorney General and the Minnesota Commissioners of Pollution Control ("MPCA") and Natural Resources ("DNR"), acting as trustees for the State of Minnesota's natural resources. On February 20, 2018, the parties entered into an Agreement and Order ("Agreement") to settle the lawsuit. That same day, the Hennepin County District Court approved the Agreement and entered it as an order of the court. The Agreement required 3M Company to pay $850 million to the State of Minnesota in the form of a restricted grant, which will be administered and implemented by MPCA and DNR."[155] As mentioned before, total assessment was $850 million which, after attorneys and other fees, the total amount decreased to $720 million.

The City of Lake Elmo v. 3M Company

Lake Elmo sued 3M twice once in 2010 and again in 2016. Several city wells have had to be shut down due to PFAS contamination. Lake Elmo sued to recover the costs of a well dug in 2007 that was unusable. The state of Minnesota has agreed to help Lake Elmo pay for a new well, with funds from the $850 million 3M settlement. To recover funds already spent, this suit, again, settled out of court, requires 3M to pay $2.7 million into the city's water account, which pays for maintaining its water system. 3M will also transfer 180 acres of farmland to the city, which the settlement says is valued at $1.8 million. Total award value: $4.5 million.

---

[155] https://www.ag.state.mn.us/Office/Cases/3M/default.asp

CHAPTER ELEVEN

The Future: Living on a Planet Steeped in PFAS

"Exposure to PFAS is an important public health concern." - Agency for Toxic Substances and Disease Registry

"Command the science" is the phrase 3M has invented and invokes when asked about its role in the global proliferation of PFAS. That is their interpretation. What seems more accurate is their attempt to "command the facts and hide them from the public". The deception, bending of facts and spinning the story in their direction is one matter but the result of those actions was the nearly five-decade long delay in dealing with or even preventing this extreme case of environmental justice. Is it too much to say that had they been honest and open in the early 1950's this crisis might have been avoided altogether? Was it too late even back then? Sadly, we'll never know. 3M chose to "command the facts" and not share them first with the residents of SoWashCo then, before they knew it, the United States and eventually the entire globe.

That misdirection was not a singular act; it was decade after decade of decisions designed to protect the company and its profits. Once the lie was told it was promulgated through subsequent generations of management and decision makers until one day, realizing the overwhelming evidence of deception meant they had to at least partially confess. Partially confess. To this day, 3M has not once admitted fault in the matter of global PFAS contamination. Even after a judgement of $850-million dollars was levied against them for PFAS-related pollution in nine SoWashCo communities did 3M admit any role in being the source. 3M is quick to deflect the conversion to what they claim is the lack of credible evidence that PFAS causes human health issues. Evidence from such notable organizations at the Centers for Disease Control (CDC)[156] to the contrary, 3M states, on a website devoted specifically to PFAS and "stewardship", "(t)he weight of scientific evidence from decades of research does not show that PFOS or PFOA causes harm in people at current or past levels"[157]. Notice they refer to past and current but spin the conversation away from the future, a future where bioaccumulation becomes more serious with each passing day. The classic quote of misdirection on 3M's part comes further down that same page when they claim "(i)t is important to note that a 'probable link' is not a 'causal link' between exposure and disease."[158] A link, however, is a link and the fact remains that, when it comes to PFAS-related health issues we just don't know beyond a shadow of a doubt the potential damage to the planet and the beings who inhabit it. We might have a better idea had 3M been honest in the early days, but it is now far too late for that. 3M wants to "command the science" and "command the conversation" by diverting attention away from the hotly debated issues of health in the hope that you won't notice this very conversation about health has been delayed since the 1950's by 3M itself.

---

[156] https://www.atsdr.cdc.gov/pfas/health-effects.html

[157] https://www.3m.com/3M/en_US/pfas-stewardship-us/health-science/

[158] https://www.3m.com/3M/en_US/pfas-stewardship-us/health-science/

The focus of this work has not been the health issues raised by PFAS; that is still an emerging and incomplete field. It is emerging and incomplete because 3M chose to remain silent on the subject of PFAS contamination for nearly fifty years. The focus here has been the tactics deployed by 3M to hide the facts and delay potential remedies for PFAS propagation. Why they lied can only be the subject of speculation. The immediate thought is to protect a highly profitable product; 3M is in business to make money after all and is accountable to its shareholders. In 2000, it was estimated that sales of PFAS-based products by 3M amounted to "about 2 percent of the company's annual sales of $16 billion (that would be $320 million)".[159] The reality now is that all of those PFAS profits down through the decades could be wiped out by past, present and future litigation for damages the substance caused and continues to cause. The consequences of those on-going decisions to hide the truth about PFAS have now come back

---

[159] Robert Bilott, *Exposure: Poisoned Water, Corporate Greed and One Lawyer's Twenty-Year Battle Against DuPont* (New York, Atria Books, 2019) pg. 98

to haunt the corridors of 3M Center. Decisions, improper and immoral decisions, made in the past have come back to sully the reputation of a once highly admired company. As PFAS accumulates in all humans and the ecosystem around the world, the role 3M played in damaging the planet has come to light.

Sadly, it doesn't appear that 3M has learned any lessons in honesty from this experience; they continue to spin the story to this day. As litigation mounts, spinning will become more and more difficult. This, then, will become a future case study in Business Schools around the world, a lesson in what not to do. Future generations will learn about this ill-formed tactic and find its folly.

The Responsible Science Policy Coalition is an organization formed in the summer of 2018. It sounds like a reputable, informed group that wholly supports responsibility in the creation of scientific policies. One of the policies this new organization appears most interested in is PFAS. "Where, then, did the Responsible Science Policy Coalition come from, and why do they care so much about PFAS?"[160] asks Michael Halpern of the PFAS Project. According to a PowerPoint presentation given at the Council of Western Attorneys General conference earlier that same year, RSPC is a new organization made up of 3M and Johnson Controls with "more members joining".[161] The last page of that presentation lists Kellerman and Heckman LLC and the Policy Navigation Group both Washington, D.C. area lobbying groups the latter of which is headquartered in a

---

[160] https://pfasproject.com/2018/10/04/what-is-the-responsible-science-policy-coalition-here-are-some-clues/
[161]

https://docs.house.gov/meetings/GO/GO28/20190910/109902/HHRG-116-GO28-20190910-SD008.pdf home in a suburb of the nation's capital. The Responsible Science Policy Coalition is nothing more than a lobbying front funded mainly by 3M to advance their agenda on a national scale. Nothing illegal about that but they are advancing 3M's agenda and not the balanced, objected, scientific information we deserve when it comes to PFAS.

As you may remember, my entire involvement with this issue arose from a college course where our small group project was to explore a topic of local environmental justice. Prompting from our professor challenged us to define the issue of PFAS and environment justice, EJ. The classic definition of traditional EJ, as offered by the Natural Resources Defense Council:

> *"Championed primarily by African-Americans, Latinos, Asians and Pacific Islanders and Native Americans, the environmental justice movement addresses a statistical fact: people who live, work and play in America's most polluted environments are commonly people of color and the poor. Environmental justice advocates have shown that this is no accident. Communities of color, which are often poor, are routinely targeted to host facilities that have negative environmental impacts -- say, a landfill, dirty industrial plant or truck depot. The statistics provide clear evidence of what the movement rightly calls 'environmental racism.'" Communities of color have been battling this injustice for decades."*[162]

But the SoWashCo story is none of those things. The average price of a home in all of Washington County is currently $320,693[163] while next door in Ramsey County,

---

[162] https://www.nrdc.org/stories/environmental-justice-movement

home of decidedly more urban St. Paul, the average home price is $256,646[164]. The United States Census Bureau pegs the median household income in Washington County at $92,376[165] while the same statistic in Ramsey County is about a third less at $62,304[166]. Finally, Ramsey County is 67.4%[167] White while Washington County is 85.9%[168] Each of those statistics for Washington County busts the myth of the traditional definition of environmental justice as defined by nearly everyone involved with that subject. Is this situation a new form of environmental justice? If the injustice is one foisted upon white, well off, well education suburbanites who own their own homes does it really count? I would argue, no, this is not a new form of EJ but rather one as old as Love Canal. The parallels between Love Canal and SoWashCo are striking: Chemical waste disposal in neighborhoods that were, at the time, decidedly not "the poor side of town". SoWashCo goes beyond the restrictions of traditional definitions in that, yes, It started out the same way Love Canal did in that a chemical company dumped its toxic waste in a residential neighborhood(s), but then PFAS and its manufacturer, ignorant of its dangers, crossed the Rubicon and created a global environmental issue the likes of which has never been seen. This is a local environmental issue that erupted across the globe. So, from that perspective, maybe SoWashCo and PFAS have redefined environmental justice or, at the very least, added an additional category. The white, well-off homeowners of South Washington County are just as much victims as are

---

[163] https://www.zillow.com/washington-county-mn/home-values/

[164] https://www.zillow.com/ramsey-county-mn/home-values/

[165]

https://www.census.gov/quickfacts/washingtoncountyminnesota

[166] https://www.census.gov/quickfacts/ramseycountyminnesota

[167] https://www.census.gov/quickfacts/ramseycountyminnesota

[168]

https://www.census.gov/quickfacts/washingtoncountyminnesota

those in urban EJ situations. Environmental justice is really injustice and injustice should be color blind. Time and environmentalists will tell if that is a true and fair statement. This issue might also be measured against the practice of "ecocide" which is defined as: Ecocide is criminalized human activity that violates the principles of environmental justice, such as causing extensive damage or destroying ecosystems or harming the health and well-being of a species including humans.[169] The key word in this definition is "criminalized" and does it apply in this case? Was 3M criminal in their activities? Improper, certainly, but criminal? Ecocide would clearly define what 3M did were it not for that one pesky word: criminalized. Since there were never any criminal charges filed against 3M in any of the dumping cases, criminalized might be considered a word too far. We are left, then, with the injustice that was heaped upon South Washington County and perhaps what was done there not only defies logic but definition as well.

---

[169] White & Heckenberg. *Green Criminology: An Introduction to the Study of Environmental Harm*, Routledge, 2014, pp 45-59.

When it comes to PFAS there is also a discernible lack of public outrage even in the affected areas. Despite Consent Orders and hundreds of millions of dollars in Settlement money people there are mostly apathetic, ill-informed or completely ignorant. Those that are aware are usually only partially so and even then, some are very misinformed or misdirected. Activism is the solution but, in many cases, that activism is thwarted by the fear of ruffling the feathers of 3M. Despite their egregious errors most local citizens overlook the company's missteps with a shrug and an "oh, well" and continue drinking, showering and cooking with a tainted water supply created by the company they choose to blindly defend. Even elected officials – or especially elected officials – seem to cower in fear of "Mother Mining". In a lengthy written response (something 3M never provided), the Congressional representative for the district that covers three of the four original dumpsites – Lake Elmo, Oakdale and Woodbury -- highlighted her positive voting record on PFAS-related legislation. She also serves as chair and vice chair of two subcommittees (appropriations on the environment and defense, respectively) where she included "$200 million in federal funding in the *Consolidated Appropriations Act, 2020* and the *Further Consolidated Appropriations Act, 2020* to begin to address these emerging contaminants"[170]. In contrast, the two United States senators from Minnesota, one of whom was running for President at the time they were contacted, have both been unresponsive to requests for information and delineation of their positions on PFAS. State senators and representatives likewise avoid the subject and contact from constituents regarding PFAS. And local mayors and city councils are a mixed bag of activism who project a "we will fight to get our share" position to those who merely dance to the dictates of 3M. Avoidance

---

[170] Letter from Betty McCollum, Member of Congress, January 17,2020

is the byword for the majority of elected officials when the subject turns to PFAS; there are few champions, advocates or activists among Minnesota's public servants.

Media has played an as yet undetermined role in getting the PFAS story out. Of course, there are social media pages and websites galore most of which promote their angle and agenda when it comes to fluorochemicals. Some of those, like the Environmental Working Group, are legitimate although EWG doesn't answer individualized question quickly, if at all. Requests for this work have gone unanswered unless prodded repeatedly. There are also more mainstream attempts at publicity such as the 2018 documentary *The Devil We Know* which focuses mainly on DuPont's role in PFAS proliferation in West Virginia and the subsequent class action lawsuit and settlement in Ohio. That story was also retold in the Hollywood film *Dark Waters* and in the book written by the litigator in the class action suit, Rob Billot, entitled *Exposure*. After all that, PFAS remains a minor story in most parts of the United States.

Like any issue of this nature, the fear is things will get worse before they get better and maybe that future gloom will be tinged with a silver lining that, once we hit rock bottom, public apathy will be replaced by public outrage and then community-based activism. As Albert Einstein once said: "Learn from yesterday, live for today, hope for tomorrow. The important thing is not to stop questioning."

We started this work with a confession, a confession that I am a retiree of the very company that committed these acts of deception. When asked how that makes me feel, what is it like to know the company where you worked for so many years did these things my answer is they made it easy for me. I read the court documents and saw with my own eyes the dates, names (some of which I recognized) and the facts they ignored. I saw how they went against the advice of their own trusted and talented employees. How year after year they didn't communicate the risks to the EPA and then, finally, there was that letter of May 15, 1998; forty-eight years, four months and five days too late. It was painful for me to see those things, to think the company I had given thirty-six years of my life to was capable of that. Painful.

I struggle with the wide range of medical facts. I struggle knowing I drank Woodbury well water for twenty-five years with no apparent affects but I struggle more with the story of the little Oakdale girl who grew up across the street from the dumpsite there and was diagnosed with Stage 4 cancer at the age of six. I struggle with an answer for how PFAS spread so widely, so quickly. I really struggle with why 3M did what it did, but I struggle most with why 3M didn't do what it should have done. There are an awful lot of I's in this paragraph but that is only because this is very, very personal to me and since 3M won't apologize for their inactions, as a thirty-six-year employee, I feel that need. I am sorry for what they didn't do and the damage their self-inflicted silence caused.

At the same time, I am hopeful. Like other daunting tasks, my innate belief in the ability of the human race as problem solvers is unsullied. We cannot change what has happened, but we can impact what will happen. Evil had a head start on us with this issue. However, the long list of wonderous works we humans have accomplished is impressive and unending. We will add the cessation of PFAS proliferation to that catalog of achievements.

If we were logical, the future would be bleak, indeed. But we are more than logical. We are human beings, and we have faith, and we have hope, and we can work.

- <u>Jacques Yves Cousteau</u>

CHAPTER TWELVE

In the News

"A well-informed citizenry is the best defense against tyranny." - Thomas Jefferson

PFAS is a volatile and controversial topic. It has that whack-a-mole characteristic: one news story appears today and as soon as that one becomes digested another story pops up tomorrow. The focus of this work was the role 3M played in creating this worldwide disaster. Yet, the subject of PFAS contamination is far from over. The issue for 3M now is the atonement for the past sins of non-disclosure; lawsuits abound and fear of the financial burden the company will need to endure in settlements and consent orders is a dark cloud looming over 3M Center. Aside from action at the state and local levels, very little has been done to address the issue legislatively. Congressional hearings have been conducted, movies made, and speeches given but little has been truly done to advance ideas of PFAS remediation and replacement.

The rest of the world is just awakening to PFAS as evident by a small sampling of news sources across the U.S.

and around the globe can attest. Here are some articles
with the citation information and a brief synopsis of content
for each. This is very much an emerging story.

**Hal Benton, "Washington state to test drinking water for
PFAS contamination linked to firefighting foam",** *Seattle
Times***, May 21, 2018**

Washington State, proactive in its hunt for PFAS,
announced that, through the state health department, 312
water systems across the state would be tested for
fluorochemical content. Tests were conducted the previous
year at Fairchild Air Force Base near Spokane. Those tests
uncovered PFAS contamination in 81 private wells in
addition to the municipal water supply in Airway Heights, a
community of 6,600. The article also discusses a federal
review that recommended new controls on PFAS some of
which are "more stringent" than present national standards.

**Carrie Fellner, "Toxic Secrets: The town that 3M built -
where kids are dying of cancer",** *Sydney Morning Herald***,
June 15, 2018**

When newspapers from 9,000 miles away come to
town to cover a story you know it is newsworthy. Tartan
High School is Oakdale's only high school. Tartan – named
after the 3M Scotch® tartan – is directly in between the
Oakdale dumpsite and the Woodbury dumpsite. Tartan has

had sixteen cancer survivors since 2002. Tartan has had four cancer deaths in seven years. These are high school aged kids who are dying of cancer, yet the Minnesota Department of Health claims Tartan's deaths do not indicate a cancer cluster.

Brady Dennis, "'Not a problem you can run away from': Communities confront the threat of unregulated chemicals in their drinking water", ***Washington Post*****, January 2, 2019**

The story focuses on the town of Parchment, Michigan. Situated in West Michigan, Parchment might be the victim of PFAS contamination from a now closed paper mill. West

Michigan in general is a PFAS hot zone due in large part to contamination by Wolverine Worldwide – makers of shoe icons like Hush Puppies®, Stride-Rite®, and Merrell® and Saucony® athletic shoes some of which are waterproofed using 3M PFAS. The article also touches on the political maneuvers and the Trump Administration's position that tougher PFAS controls would place an undue burden on water departments and small businesses.

Eric Lipton and Julie Turkewicz, "Pentagon Pushes for Weaker Standards on Chemicals Contaminating Drinking Water", ***New York Times*****, March 14, 2019**

Yes, you read that correctly: "weaker" standards. The Defense Department has been using PFAS-laden AFFF for years with the result being contaminated drinking water at nearly 200 military installations around the country. In a battleground strategy, the DoD now wants the standards lowered (or allowable parts-per-trillion increased) so they won't have to pay for expensive clean-up costs. With the EPA on one side, the Defense Department on the other and the Trump Administration somewhere in-between, the PFAS can gets kicked further down the road.

Keith Matheny, "PFAS contamination is Michigan's biggest environmental crisis in 40 years", *Detroit Free Press*, April 26, 2019

As mentioned previously, Michigan is a hotbed of PFAS contamination, so this is literally a report from the frontlines. According to this piece, there are 11,000 suspected PFAS sites around the state some testing in the 76,000 ppt range against the EPA acceptable level of 70 ppt. The focus of the article is a woman who lived directly across the street from the dumpsite used by Wolverine Worldwide where they disposed of PFAS residue left over from applying Scotchgard® to shoes. Among the lakes where a fish-eating ban is in force are Saginaw Bay on Lake Huron and all of Lake St. Clair. PFAS has reached crisis level in Michigan.

Lauren Zanolli, "Why you need to know about PFAS, the chemicals in pizza boxes and rainwear", *The Guardian*, May 23, 2019

This article answers the question posed in the headline. This brief story covers the basics from what PFAS are to what they are in, to what harm they potentially cause. Good, basic information.

Christopher Collins, "Nearly 500,000 Texans Live in Communities with Contaminated Groundwater. Lawmakers Aren't Doing Much About It.", *Texas Observer,* **June 19, 2019**

Nearly half a million Texans live within three miles of seven contaminated military installations in the state but only one of its federal congressional representatives has made any effort on establishing safer PFAS standards. And this is not small-town Texas; those bases are near cities like Dallas, Austin and San Antonio. Pollution from AFFF is so bad that residents around one of the Texas locations are taking part in a long-term CDC PFAS study.

Cheryl Hogue, "3M admits to unlawful release of PFAS", *Chemical and Engineering News,* **July 1, 2019, p. 16**

Despite a 2009 EPA order prohibiting it, the 3M manufacturing plant in Decatur, Alabama has been dumping two varieties of PFAS directly into the Tennessee River, a tributary that supplies drinking water to hundreds of thousands of people. A trade journal article, this story shows something 3M rarely does: admit they illegally dumped PFAS into a waterway.

Oliver Milman, "'Forever chemicals' found in seafood, meats and chocolate cake, FDA says", *The Guardian*, **June 3, 2019**

Do you like a nice piece of sea bass? Perhaps a medium-well ribeye is more to your liking? And who doesn't like a delicious slice of chocolate cake every now and then? According to this article nearly half of the meat and seafood tested by the U.S. Food and Drug Administration (FDA) had levels twice as high as allowable amounts of PFAS. And it gets worse: the levels in chocolate cake were 250 times higher than standards.

Dee DePass, "Turbulent terrain for 3M's CEO", *Minneapolis Star Tribune*, **September 1, 2019, p. 1D**

Redressing the sins of previous chairmen, Mike Roman now must deal on an almost daily basis with the PFAS legacy left to him by those who occupied 3M's top spot for decades. After discussing the overall business climate for the company, the article concludes with a brief discussion of PFAS and the company's looming liability.

Emily Holden, "Companies deny responsibility for toxic 'forever chemicals' contamination", *The Guardian,* **September 11, 2019**

Officials from 3M, DuPont and DuPont spinoff and now PFAS producer Chemours testifying before a Congressional Oversight Committee not surprisingly hid behind spin to deny responsibility for fluorochemical contamination while 3M's Denise Rutherford, under oath, argued the chemicals pose no human health threats at current levels and have no victims. Under oath with a straight face.

Jim Spencer, "Judge allows PFAS suit to proceed", *Minneapolis Star Tribune,* **October 1, 2019, pg. 3D**

The Ohio judge hearing this case ruled that the plaintiff, a 40-year firefighter named Kevin Hardwick, had sufficiently argued that 3M, DuPont and Chemours were responsible for the PFAS content in AFFF. The case is crucial in that it is not a class action suit which could, if decided in favor of the plaintiff, open the floodgates to innumerable lawsuits from individuals or small groups of individuals.

Jennifer Bjorhus, "Fighting chemicals in water supply", *Minneapolis Star Tribune,* **November 19, 2019, pg. 1A**

This front page story focuses on Bemidji, Minnesota where PFAS infiltration has shut down three of the city's five municipal wells. Uncertain as to the origin, city officials are looking towards another case of AFFF contamination. Since Bemidji is a long way from South Washington County, the city is looking into seeking an amendment to the 2018 settlement that would allow it to spend an estimated $16.5 million on a new and needed water treatment plant.

Jim Spencer, "Restrictions on PFAS stripped from federal bill", *Minneapolis Star Tribune*, **December 7, 2019, pg. 1D**

Seeking to minimize the potential burden of future costs, bi-partisan members of the Armed Services Committee removed proposed new PFAS restrictions from the National Defense Authorization Act. The proposed changes would have, among other items, given a "hazardous" designation to PFAS.

Daniel Ross, "Rainwater in parts of US contains high levels of PFAS chemical, says study", *The Guardian*, **December 17, 2019**

As if PFAS content in pizza boxes, chocolate cake and seafood weren't bad enough, research in the Spring and Summer of 2019 found the fluorochemicals in rainwater. Most alarming amongst the statistics? Around the Chemours plant in Wilmington, North Carolina: 500 ppt versus the allowable Federal guideline of 70 ppt. Proven fact: PFAS is now, quite literally, raining down on us.

Dee DePass, "Rash of lawsuits intensify concerns of 3M's liabilities over PFAS chemicals", *Minneapolis Star Tribune,* **December 22, 2019**

The $850 million settlement is the apex of PFAS litigation involving 3M; so far. In recapping the trail of settled and pending lawsuits, this article estimates that if and when the suing stops the total tab for 3M alone – excluding DuPont, Chemours and other minor chemical manufactures as well as their customers like the U.S. military and Wolverine Worldwide – could exceed $10 billion.

Jared Hayes and Scott Faber, "Mapping PFAS Chemical Contamination at 206 U.S. Military Sites: The Pentagon's 50-year History with PFAS Chemicals", *Environmental Working Group,* **December 31, 2019**

Pointing the finger squarely at 3M and the Pentagon, this article details the Defense Department's use and abuse of AFFF from its first application in 1967 to its proliferation at over two hundred military installations. The use of AFFF is one of the major causes of PFAS pollution and this article focuses on the armed force's role in this disaster.

Jim Spencer, "U.S. House passes bill to address PFAS; unclear if Senate will act", *Minneapolis Star Tribune,* **January 11, 2020**

A bill which the Trump Administration has already said it would veto saying it was a "litigation risk" "setting problematic precedents" and could create "unwarranted costs on the private and public sector" passed the House. The Bill, in essence, creates PFAS priorities that the EPA has failed to do. Like so much else today, the vote was along party lines and faces a very uncertain future in the Republican-controlled U.S. Senate.

Ryan Faircloth, "Woodbury takes emergency action to address water wells polluted by 3M chemicals", *Minneapolis Star Tribune,* **January 9, 2020**

With six of the City's wells closed for exceeding acceptable state levels of PFAS and a seventh well close to going over those same limits, the Woodbury City Council dipped into the SACO funds for temporary remediation uses and announced it is building a temporary water treatment plant. The City deemed it an emergency to, first of all, obtain the funds from the SACO agreement but also to get the project underway to avoid issues during the increased water usage periods in Summer.

Jennifer Bjorhus, "PFAS 'forever chemicals' found in foam in two east metro creeks", *Minneapolis Star Tribune,* **January 14, 2020**

In reporting that finding PFAS foam in two locations in the eastern portion of the Twin Cities (one location in and another adjacent to Washington County) the article sees this as an indication that the Superfund remediation process at the 3M Oakdale dumpsite is not working. Although no specific ppt numbers are mentioned, the article does say PFAS levels in the foam were "high".

Jennifer Bjorhus, "Cancer cases prompt Cottage Grove teachers to demand investigation", *Minneapolis Star Tribune,* **January 16, 2020**

Two days after the last article, the *Star Tribune* headline contains the words "Cottage Grove" and "cancer". Seven cases of cancer at Cottage Grove High School in the past six years in which all victims were under the age of 50. Thirty cases at the school since 1990, two taught in the same classroom. The Minnesota Department of Health says that's not enough evidence to call it a cancer cluster. Cottage Grove High School is a part of South Washington County Schools.

Dee DePass, "3M continues to deal with lawsuits regarding defective earplugs, PFAS chemicals", *Minneapolis Star Tribune***, January 22, 2020**

In addition to the PFAS litigation woes, 3M is also being bombarded with lawsuits surrounding a defective earplug the company sold to the U.S. military. Besides the PFAS suits, earplug cases now exceed 1,000 with more being added, quite literally, every day.

Adriana Brasileiro, "'Forever chemicals' are in nearly all U.S. drinking water, activists say. Miami is #3 on the list." *Miami Herald***, January 23, 2020**

Six out of eighty-nine wells in the Miami-Dade water supply have been closed because PFAS levels exceeded Federal standards. Miami is third on the list of locations tested in 2019 with a reading of 56.7 ppt against the federal standard of 70 ppt.

Joshua Bote, "Toxic 'forever chemicals' found in drinking water throughout US", *USA Today***, January 23, 2020**

Another take on the EWG study gives cause for concern if you live in Brunswick County, North Carolina or the Quad Cities of Iowa and Illinois. Your tap water exceeds 100 ppt against the EPA level of 70 ppt. If you live in Meridian, Mississippi your water has 0 ppt of PFAS; for now.

Dee DePass, "3M laying off 1,500 workers, reports weak financial results", *Minneapolis Star Tribune***, January 29, 2020**

Mixed in with other performance factors, the dark cloud PFAS is creating over the 3M Company has weakened their financial performance and is now having another form of impact of the company's workforce: some are being let go. Weaker than expected global results and the financial burden of PFAS litigation are cited as some of the reasons for the terminations.

David S. Cloud, "'Our voices are not being heard': Colorado town a test case for California PFAS victims", *Los Angeles Times***, January 30, 2020**

A family near a military base in a small town in Colorado is proving to be a test case for future PFAS litigation in the State of California. A new California law says that utility companies in that state must report ANY PFAS levels to their users as soon as detected. Any levels, not just those above the federal guidelines.

Tom Perkins, "The 'forever chemicals' fueling a public health crisis in drinking water", *The Guardian***, February 3, 2020**

Telling us what we already know but telling it in a concise yet all-encompassing manner, this article covers all of the places PFAS has been found from rainwater and drinking water, to sewage sludge and arctic animals. Places in placid New Hampshire where the PFAS levels are 70,000 ppt against the state's level of 12 ppt. From quiet Cottage Grove to this, a global crisis.

Dee DePas, "3M Co. agrees to pay $55M to settle PFAS contamination case in Michigan", *Minneapolis Star Tribune***, February 21, 2020**

In a first of its kind settlement, 3M has agreed to pay one of its customers – shoe manufacturer Wolverine Worldwide -- $55 million to be used for remediation purposes in the state of Michigan. Caught in the middle, Wolverine – on the very same day – signed a separate agreement with the state of Michigan to pay $69.5 million to help cover the costs of water filtration for over 1,000 homes in two communities near the Wolverine manufacturing facility. Taking its cue from 3M, Wolverine also did not admit liability.

Seventh Woodbury well has PFAS levels above water quality standards and guidelines; City of Woodbury InTouch email, February 21, 2020

The City of Woodbury announced today it has shut down the seventh of its nineteen municipal wells because PFAS levels in that well exceed allowable state and federal standards. The city's water supply has now been cut by over 35% heading into the high-demand period of summer. The previous month the City declared a water emergency, dipped into the SACO funds for temporary remediation and began construction of a temporary water treatment plant with this scenario in mind. The emergency situation anticipated by the City has come to fruition. This has not been a good news day for 3M coming the day after it had to pay $55 million dollars to Wolverine Worldwide.

Brian P. Dunleavy, "PFAS should be classified as carcinogens, researchers say", *United Press International*, March 3, 2020

Key among the findings in a report by the International Journal of Environmental Research and Public Health is a claim that "existing scientific evidence indicates that several PFAS have at least one key characteristic of carcinogens and that these chemicals should be investigated as such." Researchers conducting the study concluded PFAS need to be classified as carcinogens and treated as such.

<u>CHAPTER THIRTEEN</u>

Q & A – Tackling the Growing Tide of PFAS Questions

"A prudent question is one-half of wisdom" – Francis Bacon

When it comes to PFAS, no one has all the answers; it is the very definition of an emerging field of study. This book focused on the role 3M played in the PFAS disaster while works such as the film *Dark Waters* and the best-selling book *Exposure* were glimpses from the DuPont perspective. There are the environmental issues, health issues, remediation issues; there are issues galore. With those issues come questions and, in the case of PFAS, very few definitive answers to those questions. When the conversation turns to PFAS one of two things happen: blank stares or a barrage of questions. The public apathy, especially in those affected Minnesota communities is astounding. Whether swayed by 3M, completely disconnected or one of a myriad of other excuses, not at least being somewhat informed on the subject is irresponsible. Julie Sze, author of the book *Environmental Justice in a Moment of Danger* says a form of hegemony accounts for the lack of interest in some people. She says it's "probably akin to religion. The faith in institutions" especially ones as venerable as 3M "is (seemingly) unshakeable."

Ralph Nader once said "Nothing can stop the power of an informed citizenry when it is empowered, organized, and motivated" but we need to become informed before we can reach the far loftier goals of empowered, organized and motivated. The greater sin is to be misinformed, having a grain of truth surrounded by someone else's opinion. The uninformed, ill-informed and misinformed have no place in the battle against PFAS.

So, the first exchange needs to be the questions asked and answered. Some of those are here but, in the case of PFAS, the internet can be a useful tool when trying to become that "informed citizenry". But beware: use creditable resources not just ones that sound creditable. Stay away from social media for answers; social media is the home of bots, trolls and, ultimately, misinformation. PFAS is a serious matter that requires serious answers from serious people. Don't rely on what someone said on Facebook as your source of knowledge.

Be a member of an informed citizenry.

What follows are some typical PFAS-related questions. The answers are not solely mine; they have been gleaned from a variety of informed sources and are an amalgam of the combined global PFAS wisdom. As PFAS proliferation moved swiftly around the globe so too does this topic. Be informed but, more importantly, stay informed.

How Could 3M Let This Happen?

In the "Epilogue" there are a number of reasons given for why blaming 3M *in toto* is wrong. It was a small group within the company's siloed structure that created the problem, hid the problem, and then covered up all data pointing toward its seriousness. It was that group within the company not the company as a collective that misplaced its moral compass.

You're cooking dinner one evening when a small grease fire erupts on your stove. "I'll handle this" so you throw some water on the fire and it splashes out of the pan and quickly engulfs your cabinets and then the entire kitchen. You grab your trusty fire extinguisher, but it runs out of foam before you can cover the whole kitchen.

Spreading quickly through your home, within no time you have an inferno that your garden hose seems incapable of handling. Flames leap from your house to your neighbors and in what seems like mere seconds your entire neighborhood is ablaze.

You could have called the fire department and outside help at any point in this scenario, but you didn't. You should have realized the imminent and proven danger, but you ignored it. One mistake followed by a series of mistakes. You could have and should have been responsible, but you let ego and a misplaced sense of purpose guide the situation.

 PFAS in the early 1950's is that stovetop fire and the neighborhood is now ablaze.

What Are the Health Risks Associated With PFAS?

In sworn testimony before a Congressional subcommittee, Denise Rutherford, 3M senior vice president of corporate affairs said, "when we look at that evidence there is no cause-and-effect for adverse human health effects at the levels we are exposed to in the general population." Bear in mind, Denise Rutherford is paid by 3M, the prime prevaricator of PFAS information.

Then there is the Agency for Toxic Substances and Disease Registry (ATSDR), a federal public health agency within the United States Department of Health and Human Services. ATSDR makes the following statement:

> *The potential for health effects from PFAS in humans is not well understood. PFOS,*
>
> *PFOA, PFHxS and PFNA have generally been studied more extensively than other PFAS. In general, animal studies have found that animals exposed to PFAS at high levels resulted in changes in*

the function of the liver, thyroid, pancreas and hormone levels.[171]

The website goes on to say that PFAS may "interfere with the body's natural hormones, increase cholesterol levels, affect the immune system and increase the risk of some cancers."

This governmental agency freely admits "we don't know" all the risks so 3M and their spokesperson's flat out denial of ANY risk whatsoever has to be wrong. By making that claim, 3M is positioning itself as the keeper of knowledge greater than the Federal agency in charge of such matters. At worst the answer is the aforementioned "we don't know" but when it comes to taking an absolute stand on questions of health who would you believe? Someone who is paid to say positive things about the company that manufactured the vast majority of PFAS found presently in humans and the global ecosystem? Or the federal agency in charge of toxic substances and their associated health risks?

3M severely damaged their already tarnished reputation with Ms. Rutherford's testimony; she made herself and her company look foolish. Company before truth, self-serving spin instead of facts. If 3M cannot agree with ATSDR or any other neutral, third-party health agency that isn't in their pocket, then they are damaging even further their already sullied reputation.

Yes, PFAS can cause a variety of health issues generally in these specific areas:

1. cause developmental effects in infants

---

[171] https://www.atsdr.cdc.gov/pfas/PFAS-health-effects.html

2. lower a woman's chance of getting pregnant

3. increase a woman's blood pressure during pregnancy

4. lower infant birth weights

5. interfere with the body's natural hormones

6. increase cholesterol levels

7. affect the immune system

8. increase the risk of cancer[172]. The C-8 Science Panel concluded there was a "probable link" between PFAS and kidney and testicular cancer. Other studies have included prostate, thyroid and breast cancers as also having a probable link.

There needs to be independent research, irrefutable independent research that all sides can agree upon. This is too serious of an issue to have "our research says this" and "our research says that." We are talking about the health and well-being of planet Earth and the pettiness of denial needs to be extinguished now. We don't need to "Command the Science". What we need to do is ameliorate the science that gave us this product and use science for the greater good.

Of all the PFAS-related issues, health is the one that needs the most immediate attention. We know Ms. Rutherford and 3M to be wrong. As stated before, they hide behind the ridiculous claim that "small amounts" of PFAS are not harmful. Nearly everything in small amounts is not harmful; what a ludicrous statement theirs is. Try selling that line to the people in New Hampshire whose water supply registers 70,000 parts-per-trillion when the

---

[172] https://health.ri.gov/water/about/pfas/

acceptable level is 70 ppt. That's not a "small amount".
Neither is the 600 ppt in the water of the church down the
road from me. This is a global health concern that needs
answers and solutions. 3M needs to stop spinning and start
helping. An estimated one-quarter of the company's
business is health care related. It is beyond hypocritical to
present yourself, for profit, as a company involved in health
care then blatantly turn your back as the world's well-being
becomes threatened by one of your products. 3M has
taken a position of global irresponsibility when it comes to
helping solve PFAS-related health issues. They need to
acknowledge that health risks exist, they were the cause
and then act responsibly and help find solutions.

How Did PFAS Spread So Quickly?

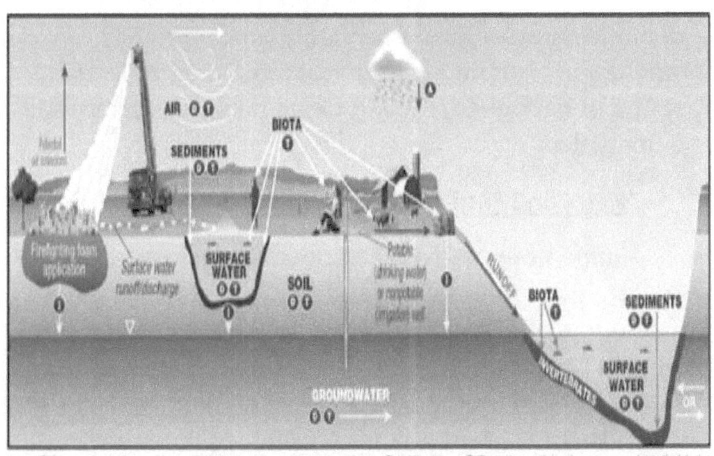

KEY ❶ Atmospheric Deposition ❷ Diffusion/Dispersion/Advection ❸ Infiltration ❹ Transformation of precursors (abiotic/biotic)

A picture is worth a thousand words or, in this case,
maybe more. If you ask the scientific minded about PFAS
they tend to promote the theory they personally believe;
that's only natural. Like so much regarding PFAS, there are
a variety of divergent opinions on its proliferation.

Water infiltration is obvious. From the groundwater contamination in SoWashCo in concert with the dumping of PFAS-tainted wastewater directly into the Mississippi and its tributaries. The Jordan Aquafer and surface water. Groundwater and surface water near every military base and airport in the world. PFAS entered the world's water supply from a variety of entry points.

There is also a group of scientists that avers to the airborne theory. They back that theory with examples like the polluted lake in Northern Minnesota out of the reach of the Mississippi and not fed from the Jordan Aquafer. How could it possibly be tainted with PFAS if not for airborne contaminants? And what about the oft-cited polar bears at the North Pole? They could have fed on tainted fish and/or been dosed with airborne PFAS. In the case of polar bears, they, sadly, could be victims of both water and airborne PFAS contamination. When the scientific community finds the definitive answer, it will probably conclude PFAS contamination was the result of a combination of factors. In the end, it happened. How it happened is a mystery still to be unraveled.

How Can I Find Non-PFAS Tainted Water?

Remember, there is no such thing as totally PFAS-free water. Even after being filtered, water will still have trace amounts of PFAS. Your goal should be to reduce PFAS-levels to amounts below allowable standards either your state's or the federal government's. With that as your objective, there are two acceptable and common methods. First, when buying bottled water read the label. You want a brand of water that at least does some form of filtration. Granular Activated Charcoal (GAC) and Reverse Osmosis (RO) are the two most common processes. One or the other listed on the label is acceptable; both processes are best. There are some brands that do both methods; in our house we drink a brand of bottled water called Deja Blue®

which is somewhat hard to find but worth the search since it uses both processes. But, again, read the label. Some bottled "spring" water has been subjected to no filtration whatsoever claiming that it comes from springs deep beneath the Earth's surface. Might I remind you that most groundwater lies deep beneath the Earth's surface.

The question you need to ask is whether it is worth the roll of the dice.

The second method would be to filter your own water at home and if you choose to go that route you should do nothing less than installing a whole-house filtration system sometimes referred to as a POET (Point of Entry Treatment) system. By only installing a filter on, say, your kitchen faucet, all the other taps in your home will still be dispensing PFAS tainted water for drinking, brushing your teeth, showering and washing your laundry. There is a word of caution: no filtration system removes 100% of PFAS in water. Whole house systems are not cheap, but they are the only way you can provide nearly PFAS-free water for your home.   Brita® Filters, to mention one brand, do not remove PFAS so don't think you're free from harm because you have an inexpensive gadget on your faucet or a pitcher in your refrigerator. Whole house GAC or whole house RO work to help remove PFAS to safe levels; accept no substitutes or other sales pitches.

Is There Any Way I Can Avoid PFAS?

No, and that is the blunt and painful truth. PFAS permeates the global ecosystem so avoiding it is already impossible. Your carpets are treated, your clothes, your dental floss and microwave popcorn bag. Your drinking water and quite possibly the very air you breathe. We live in a world immersed in PFAS. The simple answer is you can reduce your exposure, but you cannot practically eliminate all PFAS from your life. Bio-persistent, bio-accumulative and "the forever chemical". PFAS is here to stay.

Who Monitors My Tap Water for PFAS Content?

That depends on where you live. Sadly, most areas of the country are not routinely tested for PFAS content. That's one end of the spectrum. The other would be California which recently enacted legislation that requires local water authorities to notify all customers whenever any amounts of PFAS are found in the local water supply. If you have a private well, you are your own water department. You can contact a testing company to test your water or, if there is a widespread reason for concern, a local governmental agency may run those tests for you. If you are on a municipal water supply, ask. As a customer you have every right to know how often your water is tested, what it is tested for and results from the most recent tests.

What Does "Bio-Persistent" Really Mean? How Long Will PFAS Remain in Your System?

Bio-persistent is easy to define: "Tending to remain inside a biological organism, rather than being expelled or broken down". When it comes to PFAS bio-persistent means there is a tendency to, once PFAS has been absorbed into your body, to stay around. When it comes to how long PFAS will hang around let's quote the ATSDR once again since they are the federal agency in charge of toxic substances like PFAS:

> *Once in your body, perfluoroalkyls tend to remain unchanged for long periods of time. The most commonly used perfluoroalkyls (PFOA and PFOS) stay in the body for many years. It takes approximately 4 years for the level in the body to go down by half, even if no more is taken in. It appears that, in general, the shorter the carbonchain length,*

the faster the perfluoroalkyl leaves the body. Perfluoroalkyls leave the body primarily in the urine.[173]

Notice the phrase "even if no more is taken in." Not taking in any additional PFAS, at this point, is impossible since it permeates our daily lives.

Dumping Chemicals Was a Common Practice Back in the '60s So 3M Didn't Really Do Anything Illegal, Did They?

There is illegal and there is improper. 3M was not totally innocent when it dumped toxic substances directly into the ground in the 1960s. Pits at the 3M Woodbury dumpsite were supposed to be lined. They weren't. That was illegal even then. They moved forward on a number of issues completely without or with only tacit governmental approval so innocent they are not.

Beyond that is the moral obligation to be a responsible corporate citizen and endeavor to leave no trace and on that front 3M failed miserably. Their own Geology Department gave warnings. Truck drivers gave warnings. Regardless of who raised the alarm it seemed 3M ignored them.

Is There a Solution to This Predicament?

This is a war being fought on two fronts: finding a viable and effective replacement for PFAS and eliminating what is already in the environment. There is a PFAS replacement called Gen-X. It, however, is an even bigger unknown than its predecessor. Consider this from the EPA:

---

[173] https://www.atsdr.cdc.gov/toxprofiles/tp200-c1-b.pdf

> **Gen-X Chemicals**: *Gen-X is a trade name for a technology that is used to make high performance fluoropolymers (e.g., some nonstick coatings) without the use of perfluorooctanoic acid (PFOA). HFPO dimer acid and its ammonium salt are the major chemicals associated with the Gen-X technology. Gen-X chemicals have been found in surface water, groundwater, finished drinking water, rainwater, and air emissions in some areas.*[174]

Another alternative is PFBS of which the EPA says:

> **PFBS**: *PFBS is a replacement chemical for PFOS, a chemical that was voluntarily phased out by its U.S. manufacturers. PFBS has been identified in the environment and consumer products, including surface water, wastewater, drinking water, dust, carpeting and carpet cleaners, floor wax, and food packaging.*[175]

Just like the first generation chemistry, these products are being found in surface water, wastewater, drinking water in an all too familiar scenario. All this is a way of saying replacement products appear to be as questionable as the chemicals they were intended to replace.

---

[174] https://www.epa.gov/sites/production/files/2018-11/documents/factsheet_pfbs-genxtoxicity_values_11.14.2018.pdf
[175] https://www.epa.gov/sites/production/files/2018-11/documents/factsheet_pfbs-genxtoxicity_values_11.14.2018.pdf

The other front is removing it from the ecosystem which, without exaggeration, is a daunting task. You could use systems like reverse osmosis or other filtration systems being developed as effluent exits water treatment facilities before entering back into the ecosystem. But that would require cooperation of each and every water treatment facility around the world. Then there's the soil and wildlife contamination to deal with.

We seem to be inflicted with an incurable condition, but I can't think that way. My hope has been supplanted by my belief that there is a solution, an answer, a way out of this all-encompassing catastrophe. Someone out there can solve this problem. That's what we humans do, we solve problems, and this is a problem of global proportions.

CHAPTER FOURTEEN

Stumbling Toward Indecision

"I don't make jokes. I just watch the government

and report the facts." - Will Rogers

It is Tuesday April 14, 2020, 784 days since the judgement awarding $850 million to nine communities in South Washington County, Minnesota to provide clean drinking water after that essential resource was contaminated by 3M. After 112 weeks there still is no viable plan for the utilization of those funds; there still is no plan to provide uncontaminated, safe drinking water for nearly 175,000 men, women and children. In those 784 days the City of Woodbury has been forced to shut down one-third of their wells because they exceeded acceptable levels of PFAS contaminants; Cottage Grove has had to seal seventy-five percent of its wells for the same reason. In excess of two years and there is no solution, not even a hint of a solution. What <u>has</u> happened in that time is extensions and delays, talk and tests but most tragically, an inexcusable lack of urgency. This has become an exercise in how NOT to spend a billion dollars and painfully points to the folly of local government. It is Tuesday April 14, 2020 and we have just learned an initial proposed set of solutions to PFAS contamination in South Washington County – a "good, better, best case scenario" as the state refers to it -- has been pushed out again, pushed out further into the future.

A stipulation in the original settlement in February of 2018 read:

The MPCA and/or the DNR shall form a Working Group to identify and recommend projects referenced in paragraphs 14A.-C. above. Pursuant to Minn. Stat. §§ 116.155 and 115B.20, the MPCA and/or the DNR shall have the ultimate responsibility, in their discretion, to determine the projects to be implemented under this Agreement (provided that the MPCA and/or the DNR will adhere to the spending prioritizations described above). The MPCA and/or the DNR will also consult with municipalities and the Metropolitan Council as necessary and appropriate on implementation of projects under paragraph 14.A. above and may use Grant monies to reimburse those entities for projects undertaken that meet the goals set forth in paragraph 14.A. above. The MPCA and/or the DNR may use Grant monies to retain technical experts to assist the Working Group.[176]

---

[176]

https://www.ag.state.m .

One glaring omission from that stipulation was an implementation timeframe. In essence, the agreement states "here's $850 million dollars to remediate this issue but take your time, it's not like there's 1750,000 people drinking contaminated water or anything as critical as that." Point of reference: paragraph 14A mentioned in that quote says, in part:

> *The goal of this highest priority work is to ensure clean drinking water in sufficient supply to residents and businesses in the East Metropolitan Area to meet their current and future water needs.*[198]

Nothing important there, right? Just clean drinking water. Yet, here we are, more than two years after the settlement still having wells being shut down, people continuing to drink contaminated tap water and no viable solution in the foreseeable future.

Buried in the charter of one of those working groups is this mention of a timeframe, a deadline which has long since expired:

> *The MPCA and the DNR will aim to develop a conceptual plan for addressing drinking water supply with input from Subgroup 1 by the end of 2019.*[199]

Based on the information from meetings on April 14-15, 2020, that deadline is now projected to be

---

n.us/Office/Cases/3M/d ocs/Agreement.pdf [198] https://www.ag.state.m n.us/Office/Cases/3M/d ocs/Agreement.pdf
199

https://3msettlement.state.mn.us/sites/default/files/Drinking%2 0Water%20Supply%20Subgroup%20Charter.pdf

September 2020, nearly one year beyond their self-imposed completion date. It should also be noted that one of the test they are conducting that involves one of their possible solution formats is not planned to be completed before September 2021.

This is not an attempt to harpoon those individuals from the MPCA, DNR and MDH who spend their working days focused on this project. It is the stipulation itself and the system created by the settlement that is at issue. These seem to be very nice people; I have seen them in numerous meetings, and it is not my intention to impugn their reputations, credentials or efforts. It seems, or at least one gets the impression, that somehow over the years and years of litigation they never saw this coming; they never, in their wildest dreams, envisioned a day when a settlement would swing in the favor of the victims and they would need to provide a solution. I can't say it enough: these appear to be good people who have the daunting task of cleaning up a dreadful mess created by 3M. I am certain none of them were ever formally trained in how to remediate PFAS contamination in the drinking water supply of South Washington County. At the same time, however, their lack of progress does nothing but instill a lack of confidence in their ability to remediate this issue. Instead of confidence we, the victims of 3M's willful acts, see a group of bureaucrats that are overwhelmed by the task at hand. Overwhelmed and almost afraid of making a proposal, any proposal. If I was a gambling man, I would bet the house that none of the planners and project leaders responsible for this project on a day-to-day basis live in the affected area, the Hot Zone; they all live a comfortable distance away. They don't have to wonder every time they turn on their tap. They aren't forced to read bottled water labels to see if, how and when it might have been filtered. They have never opened their email to find their city has been forced to shut down yet another well due to PFAS contamination. They aren't forced to wait and wait and wait for some semblance of a solution to a decades long issue with water, the very essence of life.

It is not my intention to pillory a group of honest, hardworking individuals who are attempting to tackle an issue greater than anything else they will ever face in their careers and I certainly wouldn't do that without offering a solution. Mentioned above is the likelihood that none of the state employees involved with this project had any kind prolonged formalized training in PFAS causation and/or remediation. They are not experts. Which brings me to my suggestion: where are the experts? We know PFAS experts exist. There are even some down the road from South Washington County at the University of Minnesota. There is a myriad of subject matter experts especially on PFAS remediation issues so why haven't they been consulted? Why hasn't the MPCA or DNR contracted a true expert to help guide them to a viable, long-term and under budget solution? It has taken them over two years to get to a point of indecision when much of the ground they covered had already been pragmatically and academically explored. They have wasted two years trying to reinvent the wheel.

Instead of experts we have state bureaucrats who seem to be out of their depth, working groups that are made up of voluntary citizens with no expertise, local business and governmental leaders and in the most pitiful case of irony ever, a place for 3M itself on the aforementioned working groups. That move alone seems to give 3M *carte blanche* on how the settlement money will be spent to remediate a catastrophe they themselves created.

There is Technical Subgroup 1: Drinking Water Supply which is made up of:

The group will be composed of technical experts from MPCA, DNR, MDH, 3M, Metropolitan Council, and Washington County. The cities of Afton, Cottage Grove,Lake Elmo, Lakeland, Lakeland Shores, Maplewood, Newport, Oakdale, St. Paul Park, Woodbury, and the townships of Denmark, Grey Cloud Island and West Lakeland will each have one representative on the Subgroup. Each community, MPCA, DNR, MDH, 3M, Metropolitan Council, and Washington County can designate one alternate for when their representative is unable to participate in a meeting.

- *The Subgroup will meet at least once a month and more frequently if needed. While all members will be invited to every meeting, actual participation at a given meeting may be driven by the agenda for that meeting. It is understood that those who are interested in specific projects and approaches will be most likely to attend a particular meeting.*
- *Technical experts not affiliated with the Subgroup may be invited to consult on topics in their area of expertise.*
- *The meetings will be open to the public, and time will be reserved at the end of each meeting for public questions or comments.*[177]

Although this may sound like a group of experts (they might very well be) they just aren't experts in PFAS causation and/or remediation. These, for the most part, are state and city employees who deal with water supplies and water supply issues. To the untrained eye it seems these working groups' members do not bring the type of in-depth

---

[177] https://3msettlement.state.mn.us/technical-subgroup-1-drinking-water-supply

expertise this issue demands. They have become adept at meeting and kicking the issue down the road, but they certainly are not solution focused. This is a time for people who have spent their careers involved with per- and polyfluoroalkyl substances not the utilities manager for a small city in South Washington County. This suit and settlement are the genesis of how PFAS issues will be handled in the future. This case, this settlement that is now over two years old, is in the global spotlight. It is the first major settlement against 3M for their role in PFAS contamination. We need to move beyond the restrictions and egos of local fiefdoms. We need to be setting the global standard on how to handle the remediation of PFAS poisoning. Instead we're letting the ill-prepared local bureaucrats stumble toward indecision. We need to let the professionals handle this, but it is way too late for that; that is what coulda, shoulda and woulda happened had this been handled correctly from the start. Instead, 784 days and counting later, we still await even a proposed solution to a decades old disaster.

EPILOGUE

It Saddens Me to Say

"Man has no nobler function than to defend the truth." -
Ruth McKinney

     This has not been easy for me. Though never a rah-rah, my-company-do-or-die kind of guy there was a certain pride in working for a well-respected corporation. There was always a sense of ownership in great company achievements. It was my Oscar®, my Grammy® and my Emmy® just as much as the company's. You would always believe what the company said, announced or supported since they had never led you astray. Why would they mislead employees? There was no cause for that. It was lovingly, and perhaps a little derisively referred to as "Mother Mining"; someone who was always there to take care of employees like me. Believing in the benevolence of a corporation might seem foolish to some but by the same token there was something to be said for the consistency and reliability it endowed upon its employees. 3M was, and I would like to think, still is, that kind of company.

3M is a very siloed organization. There are line functions the names of which are usually proceeded by the word "Corporate": Corporate Accounting, Corporate Legal, Corporate Marketing and so forth. Overall, there are four business segments and then individual divisions under each segment, divisions which are focused on the segment's business sector such as health care, consumer and transportation and electronics. It was, is, and always will be easy to get lost in the structure. A division in one segment has absolutely no idea about the day-to-day business in another segment. Take it from one who has witnessed that corporate cultural diversity firsthand: five divisions in thirty-six years all different as night and day. Yes, we rolled up and reported to the Mother Ship 3M, but the day-to-day atmosphere varied wildly. We were all children of the same mother and just as you are different from your siblings, the work a day life of the Consumer Health Care Division was very unlike that of the Stationery Products Division. Different management, different objectives, different business models.

Thus, it goes without saying the business of the Commercial Chemicals Division of the '50's, '60's and beyond was totally unknown to other operating units. The bad apples responsible for PFAS contamination operated in their own world without any interference from outside their little universe.

Scandal never graced the executive suites during my thirty-six-year tenure. Of course, amongst employees there were affairs and alcoholics, divorces, DUIs and suicides all of which happened in my time there. The workforce was a combination of saints and sinners that always ended up being humans with the attending foibles and frailties. We would whisper if two divorced employees started dating; we would whisper if two single employees started dating. The whispering would reach a cacophony if two married employees appeared to be getting a little too close. We were normal humans.

There is documentary evidence in the Attorney General's database that PFAS updates were made at Board of Directors meetings beginning as early as the 1970's, but the cunning people involved in the PFAS cover-up treated the CEO and Board of Directors like everyone else blatantly telling them there was nothing to fear. By the time the CEO and Board of Directors were told of the true scope of the disaster it was already too late. They, too, were fed the same carefully crafted story; a story with just enough truth to make it not a lie.

There are those who might use the term "disgruntled ex-employee" when they cite the author of this piece. Like other times in this work, let's break that phrase down. "Disgruntled" I am not. What I am is disappointed. Disappointed in a company and its structure that allowed a small group within its corporate family to go rogue, to handle a serious issue in a wholly improper and immoral manner, a manner outside the bounds of 3M acceptability. To see those source documents, to hold them, highlight them, and finally set them down in shame is not something I was bargaining for; rather it was something done to me by those in the past who created this disaster. Those misguided employees besmirched what had been a nearly unsullied corporate history.

I am also disappointed in how the larger 3M handled the situation once it became aware of what that smaller group within had done. My plot line would have been one of honesty, self-disclosure, or voluntary remediation but my thoughts and ideas, like countless other 3M employees, were never sought. 3M chose spin over disclosure. Saying "small amounts won't hurt you" instead of saying *"mea culpa"*. Catchphrases instead of solutions. The 3M I knew and worked for all those years would have taken the higher road. Instead, they are being sued on almost a daily basis.

Finally, I am not an ex-employee; I am a retiree, and, to me, there is a monumental difference in the two. 3M and I parted ways on June 30, 2013 without animosity toward the other. I retired and did so without an ax to grind or a vision of a windmill in need of tilting.

And things would have remained that way if a fellow classmate had never said "What do you think about doing something around PFAS contamination in South Washington County?"

In her 2017 essay "Tell Me How It Ends: An Essay in 40 Questions", Valeria Luiselli says "it is never inspiration that drives you to tell a story, but rather a combination of anger and clarity"[178]. That was never truer than in my relationship with this story. The anger in me is generated from the betrayal left by an organization that dominated my life for decades and its offensive actions. The clarity derives from the searing truth in page after page of internal documents for all to see in court records. Anger and clarity did indeed tell this story.

This is an issue that needs a Greta Thunberg. Maybe that person will be Mark Ruffalo since he has stepped to the forefront with the film *Dark Waters*. But that movie leaves 3M unscathed and that is not what happened in real life. 3M is just as if not more so culpable as DuPont for the proliferation of PFAS. The "In the News"

---

[178] Luiselli, Valeria. *Tell Me How It Ends: An Essay in 40 Questions.* Coffee House Press, 2017.

chapter shows that attempts are being made to drive public outrage yet there seems to be only small pockets of resistance.

My hope is in time that will change.

As the case filed by the State of Minnesota against 3M was progressing, the Attorney General's office, as is done with most large-scale litigation, periodically tested their case before mock juries. In an interview, Lori Swanson, the then Attorney General said as the staff in the AG's office presented their case to those mock jurors they first became angry and then, invariably, that anger turned into terror, sheer fear of what PFAS contamination meant to them, their families and the planet they inhabit. Anger, once again, a result of 3M Company's role in PFAS proliferation.

Back to the beginning: this has not been easy for me. A great portion of this work is cited from source 3M documents and therein lies the rub: it is fact through the company's eyes which, many times, differed from substantiated fact. Yet there were also times when the truth within the documents was not what was revealed to the public at large. To read a document was like seeing a snapshot in time; being blessed with the capability of following the documentation paper trail made a significant difference. In each of the cases, documents would start out firmly encased in fact, in the truth, but over time the facts would bend and twist until they had morphed into something only vaguely resembling their starting point. It was then you returned to that single starting point to compare and contrast the alpha and omega: the fact and the spin.

That very same spin was created by a few to protect a highly profitable product line. Not always lies, per se, but many times not quite the "whole truth" either. There was, after all, millions if not billions of corporate profit dollars to protect. As with the global population, tens of thousands of 3M employees never knew what was going on as we too were never really told; the real truth, that whole truth was kept among the few. Blaming all of 3M for the mistakes of a few is unwarranted. However, to this day 3M resorts to spin with its employees still not quite disclosing the whole truth only revealing a little truth at a time. There were only a few bad apples who made poor choices and ruined the barrel. In the end, the buck needs to stop somewhere. The 3Mers of today are left to atone for the sins of their forbearers.

That's not fair but, then again, neither is contaminating the planet.

Lakota legend says pure spring drinking water is brought to the surface by a living force or spirit called *Wiwila*. Whenever a Lakota would take a drink from a spring, any spring, anywhere they would say: "Wiwila, have pity on me as I drink." The world has gotten to a point where the wisdom of the indigenous needs to be invoked once more. The next time you turn on your tap, look to the ancient powers and ask "Wiwila, have pity on me as I drink."

"Three things cannot be hidden long: the sun, the moon and the truth"

-   Buddha

DISCLAIMER

There have been numerous attempts to directly engage 3M in this work. My first request for information was made on September 27, 2019; the reply from the External Communications Manager at 3M to that initial email was sent on November 4, 2019 over seven weeks later. That reply from 3M asked questions about who I was talking to and what my intentions were regarding the use of the information I was requesting.

My response to those questions, one week later on November 11, 2019, was a nonresponse; 3M is a public company and I am myself a shareholder so answering questions about how I, as a stakeholder in the company, was planning to use public information seemed inappropriate. I did not answer their question any more than they did not answer mine. More than three weeks later on December 4, 2019 I received a reply from a "Media Relations Specialist" thereby making it obvious I would no longer be dealing with management at 3M. In a rapid-fire exchange of emails, he was very eager to "hop on a quick call" rather than answering my questions in writing. Our email conversation ended that same day when it became a surefire certainty I was not going to receive answers to my questions in writing and 3M's preferred method of communication was one where the worst-case scenario would be one of "he said, she said". A local reporter revealed to me they had encountered the same issues when attempting to gather information from 3M for a PFAS-related story. "Phones were never answered; it was always voice mail and never once did I receive a written response to questions posed in an email." 3M continues to "Command the Conversation".

Therefore, 3M's position on topics for this work have been gathered from discovery documents, original source documents, media coverage and 3M's own website. This is a fact-based work and all information from outside sources is cited. My hope was that I would be able to ask questions and receive answers in an open and honest exchange with 3M, but it became clear the only way that was going to happen would be if they would be able to control the conversation; with multiple lawsuits and what one can only imagine hundreds if not thousands of requests for information from sources around the world, 3M has obviously adopted a defensive posture and prioritization of responses is paramount. Those cases of unresponsiveness only taint an already damaged image even further.

<u>**APPENDIX A**</u>

The Proliferation of PFAS – A Timeline of a Dozen Key Events

When the discussion creeps toward the key events that write the biography of PFAS the lens you use to view those events alters not only the perception, but which events come into focus and which blur from that chosen perspective. The lens of health issues sees different dates more clearly than the lens of DuPont or the lens of eco-proliferation. Since this work focused on the role 3M Company played that lens sees the dozen dates listed below more sharply than the others.

1938	DuPont chemists accidently discover what would become fluorochemicals.
1947	Construction begins on 3M Chemolite, Cottage Grove, MN

1950	Process patent issued to 3M from mass production of fluorochemicals; first toxicity test kills lab mice.
1958	Minnesota Department of Health conducts tests on fish in Cottage Grove with effluence from 3M; fish die within five minutes.
1963	3M and the Department of Defense begin work on aqueous film forming foam (AFFF).
1966	Woodbury dumpsite is closed after chemicals found in neighboring wells.
1975	Outside researchers tell 3M they've found PFAS in blood samples from parts of the country outside Minnesota.
1983	Oakdale dumpsite is so polluted with a cocktail of chemicals (including PFAS waste) that its clean up qualifies as a Superfund site.
1998	3M sends a letter to the EPA for the first time identifying PFAS as a toxic substance.
2000	3M announces to the general public a phase out of "perfluorooctanyl chemistry used to produce certain repellents and surfactant products".
2002	3M replaces its old C-8 chain perfluorooctanyl chemistry with a C-4 chain making "the forever" chemical half as dangerous.
2018	The State of Minnesota and 3M settle a lawsuit filed on behalf of nine communities in South Washington County for a record $850 million.

BIBLIOGRAPHY

Print Media

David Barboza, "E.P.A. Says it Pressed 3M for Action on Scotchgard Chemical", *New York Times*, May 19, 2000

James Berman, "The Three Essential Warren Buffett Quotes to Live By", *Forbes*, April 20, 2014

Robert Bilott, *Exposure: Poisoned Water, Corporate Greed and One Lawyer's Twenty-Year Battle Against DuPont* (New York, Atria Books, 2019)

Henry Wadsworth Longfellow, *Song of Hiawatha*, (Boston, Ticknor & Fields, 1855)

Letter from Betty McCollum, Member of Congress, January 17,2020

The Multiple Risk Factor Intervention Trial (MRFIT). A national study of primary prevention of coronary heart disease. (1976)

Nathaniel Rich, "The Lawyer Who Became DuPont's Worst Nightmare", *New York Times Magazine*, January 6, 2016

Julie Sze, Environmental Justice in a Moment of Danger (Oakland, CA University of California Press, 2020)

White & Heckenberg. *Green Criminology: An Introduction to the Study of Environmental Harm*, Routledge, 2014, pp 45-59

Websites

- https://msdsauthoring.com/msds-safety-data-sheet-chemicals-osha-msds-rules
- https://en.wikipedia.org/wiki/Effluent
- https://www.google.com/search?ei=Rjn-XdKHJcWwtAa8anIBg&q=oervical+dislocation&oq=oervical+dislocation&gs_l=psyab.3..0i13l10.96131.96131..98454...0.3..0.80.80.1......0....2j1..gws-wiz.......0i71.cr6GFrBmUoA&ved=0ahUKEwj SwtarhsfmAhVFGM0KHbx8CmkQ4dUDC As&uact=5
- https://www.dictionary.com/browse/in-extremis?s=t
- https://www.cityrating.com/crime-statistics/minnesota/woodbury.html
- https://money.com/collection/2018-best-places-to-live/5361448/woodbury-minnesota-2-2/
- https://www.homesnacks.net/best-cities-for-families-in-minnesota-1211110/
- https://population.us/mn/woodbury/
- http://worldpopulationreview.com/us-cities/woodbury-mn-population/ • https://www.google.com/

<u>Websites (Cont.)</u>

- https://www.ewg.org/interactivemaps/2019_pfas_contamination/map/?_ga=2.83688918.1350710445.1579527447159866873.1575744099
- https://www.michigan.gov/pfasresponse/0,9038,7-365-86511_82704_83952---,00.html
- https://www.easycalculation.com/unit-conversion/Parts_Per_Trillion-pptParts_Per_Billion-ppb.html
- https://www.easycalculation.com/unit-conversion/Parts_Per_Trillion-pptParts_Per_Million-ppm.html
- https://ohiovalleyresource.org/2016/10/21/chemicals-found-water-means/
- Section 8(e), the substantial risk information reporting provision of the Toxic Substances Control Act (TSCA). -_https://www.epa.gov/assessing-and-managingchemicals-under-tsca/tsca-section-8e-reporting-guide
- https://www.investopedia.com › terms › de-novo-judicial-review
- https://www.epa.gov/enforcement/3m-company-settlement
- https://www.bizjournals.com/twincities/news/2017/07/10/3m-buys-naming-rights-tomariucci-arena.html
- https://www.bodyshopbusiness.com/3m-automotive-aftermarket-division-keepsnascar-teams-fast-lane/
- https://www.latimes.com/archives/la-xpm-1985-11-26-fi-2140-story.htm
- https://www.atsdr.cdc.gov/pfas/health-effects.html

- https://www.3m.com/3M/en_US/pfas-stewardship-us/health-science/
- https://www.nrdc.org/stories/environmental-justice-movement
- https://www.zillow.com/washington-county-mn/home-values/
- https://www.zillow.com/ramsey-county-mn/home-values/
- https://www.census.gov/quickfacts/washingtoncountyminnesota
- https://www.census.gov/quickfacts/ramseycountyminnesota
- https://www.atsdr.cdc.gov/pfas/PFAS-health-effects.html
- https://health.ri.gov/water/about/pfas/
- https://www.jdsupra.com/legalnews/state-by-state-regulation-of-per-and-82542/
- https://www.atsdr.cdc.gov/toxprofiles/tp200-c1-b.pdf
- https://www.epa.gov/sites/production/files/2018-11/documents/factsheet_pfbs-genxtoxicity_values_11.14.2018.pdf

Minnesota Attorney General's Office

https://www.ag.state.mn.us/Office/Cases/3M/StatesExhibits.asp

- Exhibit 1001, State of Minnesota v. 3M Co., Court File No. 27-CV-10-28862
- Exhibit 1009, State of Minnesota v. 3M Co., Court File No. 27-CV-10-28862
- Exhibit 1020, State of Minnesota v. 3M Co., Court File No. 27-CV-10-28862

- Exhibit 1025, State of Minnesota v. 3M Co., Court File No. 27-CV-10-28862
- Exhibit 1043, State of Minnesota v. 3M Co., Court File No. 27-CV-10-28862
- Exhibit 1046, State of Minnesota v. 3M Co., Court File No. 27-CV-10-28862
- Exhibit 1058, State of Minnesota v. 3M Co., Court File No. 27-CV-10-28862
- Exhibit 1065, State of Minnesota v. 3M Co., Court File No. 27-CV-10-28862
- Exhibit 1068, State of Minnesota v. 3M Co., Court File No. 27-CV-10-28862
- Exhibit 1083, State of Minnesota v. 3M Co., Court File No. 27-CV-10-28862
- Exhibit 1118, State of Minnesota v. 3M Co., Court File No. 27-CV-10-28862
- Exhibit 1191, State of Minnesota v. 3M Co., Court File No. 27-CV-10-28862
- Exhibit 1195, State of Minnesota v. 3M Co., Court File No. 27-CV-10-28862
- Exhibit 1252, State of Minnesota v. 3M Co., Court File No. 27-CV-10-28862
- Exhibit 1254, State of Minnesota v. 3M Co., Court File No. 27-CV-10-28862
- Exhibit 1471, State of Minnesota v. 3M Co., Court File No. 27-CV-10-28862
- Exhibit 1479, State of Minnesota v. 3M Co., Court File No. 27-CV-10-28862
- Exhibit 1488, State of Minnesota v. 3M Co., Court File No. 27-CV-10-28862
- Exhibit 1496, State of Minnesota v. 3M Co., Court File No. 27-CV-10-28862

- Exhibit 1502, State of Minnesota v. 3M Co., Court File No. 27-CV-10-28862
- Exhibit 1664, State of Minnesota v. 3M Co., Court File No. 27-CV-10-28862
- Exhibit 1690, State of Minnesota v. 3M Co., Court File No. 27-CV-10-28862
- Exhibit 1872, State of Minnesota v. 3M Co., Court File No. 27-CV-10-28862
- Exhibit 2589, State of Minnesota v. 3M Co., Court File No. 27-CV-10-28862
- Exhibit 2602, State of Minnesota v. 3M Co., Court File No. 27-CV-10-28862
- Exhibit 2621, State of Minnesota v. 3M Co., Court File No. 27-CV-10-28862
- Exhibit 2639, State of Minnesota v. 3M Co., Court File No. 27-CV-10-28862
- Exhibit 2640, State of Minnesota v. 3M Co., Court File No. 27-CV-10-28862
- Exhibit 2660, State of Minnesota v. 3M Co., Court File No. 27-CV-10-28862
- Exhibit 2662, State of Minnesota v. 3M Co., Court File No. 27-CV-10-28862
- Exhibit 2695, State of Minnesota v. 3M Co., Court File No. 27-CV-10-28862
- Exhibit 2719, State of Minnesota v. 3M Co., Court File No. 27-CV-10-28862

Documentary Films

The Devil We Know, Stephanie Soechtig, Jeremy Seifert - Directors, Joshua Kunau – Producer, 2018, YouTube.